GANDHI
on WAR
and PEACE

GANDHI
on WAR
and PEACE

Rashmi-Sudha Puri

PRAEGER

New York
Westport, Connecticut
London

Library of Congress Cataloging-in-Publication Data

Puri, Rashmi-Sudha.
  Gandhi on war and peace.

  Bibliography: p.
  Includes index.
  1. Gandhi, Mahatma, 1869-1948 – Views on war.
2. Gandhi, Mahatma, 1869-1948 – Views on peace.
3. War. 4. Peace. I. Title.
DS481.G3P85 1986   954.03′5′0924   86-20489
ISBN 0-275-92303-7 (alk. paper)

Library of Congress Catalog Card Number: 86-20489
ISBN: 0-275-92303-7

First published in 1987

Praeger Publishers, 521 Fifth Avenue, New York, NY 10175
A division of Greenwood Press, Inc.

Printed in the United States of America
∞

The paper used in this book complies with the Permanent
Paper Standard issued by the National Information Standards
Organization (Z39.48-1984).

10 9 8 7 6 5 4 3 2 1

To my parents

in eternal gratitude
for their boundless love and generosity of kindness

FOREWORD

Mahatma Gandhi coined the term nonviolence and championed the cause of peace in an age that has shamelessly surpassed earlier centuries in its slavish addiction to war, cruelty, and bloodshed. While he pioneered new modes of nonviolent resistance and profited by his traumatic experience of ambulance work in three wars, Gandhi evolved a coherent political philosophy that unconditionally repudiated any rationalization of war or any ethical justification of systematic violence. His theoretical standpoint was sufficiently subtle and complex to be misunderstood by many, but even more puzzling to his myriad admirers was his lifelong insistence on the rightness of his indirect participation in the Boer War, the Zulu Rebellion, and World War I. The latter theme deserves consideration, which is now available in this careful and conscientious study by Dr. Rashmi-Sudha Puri.

Dr. Puri conducts her investigation with the clear recognition that Gandhi never wavered in holding the issue of war to be supremely important, posing hard questions for all those concerned with the morality, feasibility, and corrupting consequences of violence as an instrument of conflict resolution. She has provided an accurate, well-documented survey of Gandhi's policies and precepts in relation to war and peace throughout his long and eventful career in South Africa and India. After examining his direct involvement in the Boer War and the Zulu Rebellion, she considers the difficult choices Gandhi had to make on the advent of World War I and even more in launching satyagraha (nonviolent resistance) amid British preoccupations with World War II. By juxtaposing Gandhi's unorthodox reflections on war with his shrewd observations on the problem of peacemaking, and stressing his global perspective on self-determination and mutual cooperation, Dr. Puri presents a balanced and convincing account of Gandhi's varying applications and formulations of an essentially consistent, though constantly evolving, standpoint.

Dr. Puri's historical and sociological assessment does not fall prey to the familiar temptation of explaining away an exacting ethical and political creed by absolutizing a specific social context or by confounding tactical constraints with theoretical limitations. To do so is to diminish the impact of a daring thinker and doer without in any way enlightening the reader. Dr. Puri does not attempt a rigorous analysis of concepts or a definitive interpretation of events, but she succeeds in setting forth a vast and impressive array of evidence in support of the experimental and evolutionary nature of Gandhi's persisting attempts to fuse theory and practice, showing flexibility without expediency, firmness without insensitivity, and fidelity without self-deception.

Dr. Puri's work is both judicious and thought-provoking, a valuable addition to the prolific literature on Gandhi, which can also aid all those who wish to explore the issues of war and peace in the context of the twentieth century, with its staggering unfinished business in securing the minimal conditions for global stability and social justice, survival without endemic exploitation, and effective cooperation toward a progressive redistribution of power, wealth, and human energy.

Raghavan Iyer
Santa Barbara

PREFACE

Professor Nirmal Kumar Bose, a keen student and a close companion of Gandhi in the years immediately preceding his assassination, at a seminar held in the Gandhi Centenary Year (1969) at the Indian Institute of Advanced Study, Simla, observed:

> Mahatma Gandhi looked upon the problem of War as the most important problem which faced the contemporary world.

> He believed that unless we were able to devise a method of collective action which would be more efficient than war in the resolution of conflicts, humanity would be involved in a race for armaments which might prove disastrous and suicidal. . . .

And yet, curiously enough, not much work on Gandhi's concern for peace and abhorrence of war seems to have been done. The somewhat presumptuous tendency seems to be that all has been said and done when Gandhi's contribution to India's emancipation or his nonviolence and satyagraha (nonviolent resistance) have been elaborated, discussed, or eulogized. Amazingly, in the growing mass of Gandhiana, his overall creed of nonviolence is supposed to reveal Gandhi's view of the world and of interstate relations, his design for transforming humanity by inculcating appropriate values and taking tangible, specific measures that will pave the way to a more humane, sane, and just world order, without taking a closer look at what he had to say on these crucial matters. Comments and assessments regarding Gandhi's perception of the world, in sharp contrast to the massive literature on his other aspects, are quite sparse. Additionally, these few works take a casual, limited, or opinionated view of the subject. Preoccupied as this small number of works is with some other aspect of Gandhi's grand personality, or oceanic work, or myriad concerns, it perforce virtually overlooks his deeply deliberated, closely reasoned, and consistently expressed views on the vital issues of war and peace in the world.

The very thought of viewing Gandhi, "the prophet with a universal message," only in terms of the history and culture of India is most irreverent and unfair, to say the least. Without being a public relations agent, it should be possible to put academic scrutiny Yugapurush Gandhi's deliberated

ix

thinking, views, and convictions on issues of universal concern. It is fairly well established now that Gandhi was not given much to rhetoric that pretends to put forth a view without indicating a clear commitment to it. His ideas on war and peace, accordingly, are eminently challenging for academic analysis.

Here is a subject that does not admit of oral evidence, for whatever Gandhi had to say concerning the world--its problems and prospects--could not possibly have been whispered into private ears, howsoever close they might have been to him. Even if it had been, the recipients of his opinions would never have kept them to themselves but would have made them known through the printed word to the world for which his message was meant.

Professor K. P. Karunakaran interested me in this eminently valuable and fascinating subject for research, which duly earned me the Ph.D. degree from the Panjab University. My subsequent study and teaching two summers at the Portland State University now find shape in the present book, which I hope will, to an extent, fill the gap in the literature on the subject. If it serves, howsoever modestly, to induce understanding of Gandhi and spread his concern for moral regeneration of peoples leading to a kinder world to live in, I shall feel well rewarded.

Rashmi-Sudha Puri

Panjab University

ACKNOWLEDGMENTS

I have benefited greatly from a large number of friends too numerous to single out. Still I would like to gratefully acknowledge my indebtedness specifically to my teacher, Professor Iqbal Nath Chaudhry, Professors K. P. Karunakaran, Ashin Das Gupta, Shadi Lal Malhotra, Randhir Singh, Shri T. K. Mahadevan, Shri Vijay Kataria, and Dr. Manohar Lal Sharma.

Joan Bondurant, a serious scholar of Gandhi, read the work at an early stage with sensitivity and made some enlightened comments of a fundamental nature, which now find their due place in the text. To her, as indeed to Professor Usha Mehta, a disciple of the Mahatma, I am forever beholden.

The many kindnesses of Dr. David Taylor, who painstakingly went through the manuscript and offered valuable suggestions, and of Arthur and Fiona Bourne whose encouragement was tangible and timely, are warmly cherished.

The late Prime Minister Shrimati Indira Gandhi, during a visit to my area, kindly spared some time to talk to me on the mahatma she knew; I am grateful for the cherished memory of the talk.

I am greatly obliged to Professor Raghavan Iyer for his gracious foreword to this book.

Thanks are also due to the staff of the libraries at Panjab University, the Gandhi Smarak Nidhi, and the Nehru Museum at New Delhi for their kind assistance. I am obliged for the implicit permission to cite from copyright material. For the interpretation and the opinion expressed in the work, however, I alone am responsible.

My young daughter Suparna and son Shantanu had to go without a mother's attention during my labors on this work. They acquiesced with the touching grace of innocent children, which fills me with pride and emotion. My Biji gave them the love I could not and provided constant, concerned, and loving encouragement in my research. To her my gratitude is deep and eternal.

To my father and my husband, my gratitude is personal and profound. But for their kind encouragement and understanding insistence, this work would not have seen the light of day.

CONTENTS

GANDHI
on WAR
and PEACE

1
THE SHAPING OF A MAHATMA

Mohandas Karamchand Gandhi never presumed any originality to his theories: "I do not claim to have originated any new principle or doctrine," he maintained; "Truth and nonviolence are as old as the hills."[1] Out of modesty or by sheer force of habit to speak the truth, Gandhi generously and repeatedly acknowledged the men and ideas that influenced his thinking and actions. Quite early in his career he declared, "whilst the views expressed in Hind Swaraj are held by me, I have but endeavoured humbly to follow Tolstoy, Emerson and other writers, besides the masters of Indian Philosophy."[2] And to support this view, he appended a list of works of these writers to his work, Hind Swaraj.[3] It would, however bear analysis to take stock of the facts in this regard.

The earliest influences upon Gandhi, as he tirelessly repeats, were his parents, who were devout followers of the cult of Vishnu. "The outstanding impression my mother has left on my memory," he holds, "is that of saintliness."[4] It seems that as a boy he learned his first lesson in ahimsa (nonviolence) also from his father. A remarkable chapter in Gandhi's Autobiography describes his father's reaction to the young Mohandas's confession of having committed theft: "This was, for me, an object-lesson in Ahimsa. Then I could read in it nothing more than a father's love, but today I know it was pure Ahimsa."[5]

According to Gandhi his predilection toward what he called "disciplinary resolutions" was also the result of his mother's saintliness and her influences upon his earliest awareness. Just a few days before his death, Gandhi told this to Vincent Sheean, who duly recorded it in Lead Kindly Light, an intensely personal record of his short experience of Gandhi in January 1948.[6]

1

Doubtlessly, Gandhi owed a great deal to his mother. The triple vow of abstaining from wine, women, and meat, which Gandhi observed all his life, was imparted to him earnestly by his mother on the eve of his departure to London in the summer of 1888. He was neither conscious nor did he realize the full implications of these early influences at the time, of course. But the parental influence most certainly played an important, if not decisive, role in the formation of Gandhi's ideas. His desperate search in London for a vegetarian restaurant clearly illustrates this fact. When he arrived in London, he was a vegetarian not by conviction but because his mother had insisted that he remain one. His discovery of a vegetarian restaurant is significant for he came in contact with a British vegetarian group and read literature on vegetarianism, such as Henry Salt's Plea For Vegetarianism and Howard Williams' The Ethics of Diet. All this lent conviction to Gandhi's belief and made him determined to stay vegetarian for he had discovered the ethical message of vegetarianism. "From the date of reading this [Salt's] book I may claim to have become a vegetarian by choice," maintains Gandhi.

> I blessed the day on which I had taken the vow before my mother. I had all along abstained from meat in the interests of truth and of the vow I had taken, but had wished at the same time that every Indian should be a meat-eater, and had looked forward to being one myself freely and openly someday, and to enlisting others in the cause. The choice was now made in favour of vegetarianism, the spread of which henceforward became my mission.[7]

His contact with the vegetarian club created in Gandhi a fresh awareness of his Indian origin. Till then, he had paid little attention to the scripture loved by his parents, the Bhagwad Gita. It was toward the end of his second year in London that he first acquainted himself with the Gita, given to him by two Theosophist brothers, translated into English as The Song Celestial by Edwin Arnold; he also read Lord Buddha's teachings in Arnold's The Light of Asia.

By no means well read at the time, Gandhi was not much inclined toward what passes today as intellectual pursuits.[8] Thus, in London he was not drawn to men of science, rationalists, Fabian socialists, or to intellectuals. Basically, he was shy and of a rather nervous temperament. So he stuck to the vegetarian club and the Theosophist circle, where he began to take an active interest in religion. Accordingly, he acquainted himself not only with Hinduism but with other religions as well. He read the Bible for the first time; the

Old Testament did not appeal to him much, but he went through the New Testament with great interest. The "Sermon on the Mount" in particular left a deep impression upon him: "But I say unto you, that ye resist not evil" touched him tremendously. His young mind tentatively set about to synthesize the teachings of the Gita, The Light of Asia, and the Sermon on the Mount. For Gandhi the idea of renunciation was the unifying principle in these three. And so, at the conclusion of his three years of legal studies, he "left England an augmented Indian and that means a stronger man as well as his mother's son."[9] The three years in England had been important but not entirely decisive in the development of his mind.

Gandhi acknowledges, "Three moderns have left a deep impress on my life and captivated me: Raychandbhai by his living contact; Tolstoy by his book, The Kingdom of God is Within You; Ruskin by his Unto This Last."[10]

He met the 25-year-old Gujarati poet-saint Rajchandra Ravjibhai Mehta, whom Gandhi refers to as Raychand, on his return to India in 1891. Although Gandhi's actual personal contact with Raychand exceeded hardly two years, for Gandhi went to South Africa in 1893, the extensive learning and extraordinary lucidity of the poet left an indelible mark on him. Even in such a limited span of time, a very intimate and thorough exchange of thought occurred between the two young men. Gandhi developed tremendous faith in Raychand and was convinced that the latter would never lead, or allow him to be led, astray. At this stage in Gandhi's life, Raychand truly became his friend, philosopher, and guide: "In my moments of spiritual crisis . . . he was my refuge."[11] Even while in South Africa, in moments of stress he would write to Raychand and solicit his guidance. For instance, just before the poet's death in 1901, Gandhi found himself faced with a spiritual crisis. His Christian friends pressed him to convert to Christianity. Under the circumstances, Gandhi had no one to look to but Raychandbhai and addressed about 27 questions on the attainment of moksha and various religions in a letter to him.[12] Raychandbhai, in response, counseled Gandhi not only to adhere to his own faith firmly, but also to respect other faiths, and in due course transcend all faiths in pursuit of the ultimate goal, truth. It must be remembered that Raychandbhai was a Sthanakavasi Jain, not a doctrinaire Jain but one who in several ways combined elements from both Vaishnava and Jain persuasions. And so, Vaishnavite Hinduism, liberal Jainism, and veneration for the personality of Jesus Christ all combined to form the basis of Gandhi's moral fiber.

Of the Western writers to whom Gandhi acknowledges profound indebtedness, Leo Tolstoy is foremost--and for good reasons. Tolstoy was the only eminent Western writer of interest to Gandhi who was Gandhi's contemporary; Gandhi had read his works and through correspondence had also established some personal contact with him. Gandhi assigned Tolstoy a place of highest prominence among all the Western writers who he said had impressed him.[13] Let us examine the impact of Tolstoy first since Gandhi read him first and then Ruskin.

According to Gandhi, he was introduced to the concept of "bread labor" by Tolstoy. The originator of the concept, incidentally, was not Tolstoy but another Russian by the name of Bondaref. Gandhi incorporated the idea of bread labor after Ruskin's Unto This Last had moved him immensely. He seems to have familiarized himself with the works of Tolstoy even as a student in England. His earliest reference to the celebrated Russian appears in the Guide To London, which Gandhi started writing in England and completed in 1893 as he was leaving India for South Africa. However, the particular work of Tolstoy that left an abiding imprint on Gandhi was read by him in the very first year (1893) of his stay in South Africa. "Tolstoy's Kingdom of God is Within You," records Gandhi in his Autobiography, "overwhelmed me. It left an abiding impression on me. Before the independent thinking, profound morality and the truthfulness of this book, all the books given to me . . . seemed to pale into insignificance."[14]

Gandhi's declaration that he learned the lesson of ahimsa from his wife is well known.[15] Yet he also stresses repeatedly that it was Tolstoy who converted him to it. Recalling the impact that Kingdom of God had had on him, on the occasion of the Birth Centenary Celebrations of Tolstoy (September 10, 1928), Gandhi said:

> I read the book forty years ago.*
> At the same time I was sceptical about many things and sometimes entertained atheistic ideas. When I went to England, I was a votary of violence, I had faith in it and none in non-violence. After I read this book, that lack of faith in non-violence vanished. Later, I read some of his other books but I cannot describe what effect they had on me. I can

---

*Gandhi's memory seems to have played him a trick here. He could not possibly have read the book 40 years ago, that is, in 1888 when he had just landed in England. In fact, he could not have read the book in England at all, since he left for India toward the end of 1891 and the book actually came into print in 1893. Moreover, his Autobiography suggests (on page 99) that he could not start reading until 1894 when he was in South Africa. Perhaps he was being rhetorical.

only say what effect his life as a whole had
on me.16 [Emphasis added]

Gandhi repeatedly acknowledged the debt, stating in
categorical terms the transforming effect that Tolstoy's books
had on his inner, spiritual growth. Addressing a meeting at
the Germiston Literary and Debating Society on "The Ethics
of Passive Resistance," Gandhi said that Tolstoy had saved
him from skepticism, had helped him define passive resistance,
and claimed that Tolstoy was "the best and the brightest
modern exponent of the doctrine."17 So boundless was his
admiration for Tolstoy that he would make rather sweeping
statements: "I know no author in the West who has written
as much and as effectively for the cause of non-violence as
Tolstoy has done."18 He even went to the extent of maintaining that he knew of no one in India, or elsewhere, who had
as profound an understanding of the nature of nonviolence
as Tolstoy had, and who tried to follow it as sincerely as
Tolstoy did.

As we noted above, Gandhi was influenced more by Tolstoy's life rather than by his works. In Gandhi's opinion,
what Tolstoy said or wrote was not entirely new,19 and yet
there was magic in Tolstoy's words for he acted upon what
he preached. In the matter of nonviolence, Gandhi wanted
his countrymen to follow Tolstoy rather than the Indian saints.
The youth of the country was called upon to "learn the lesson
of self-control from Tolstoy's life."20 Gandhi felt that Tolstoy's simplicity was too transparent to be missed; indeed, it
was extraordinary in that it was not only manifest outwardly
but was deeply ingrained within him. One of the aristocracy,
having all the wealth and luxury at his disposal, Tolstoy dis-
carded it once the sheer futility of it all had dawned upon
him. Gandhi, himself veering toward the force of renunciation,
quite naturally admired and emphasized the greatness of Tolstoy as a man, for with Tolstoy to believe was to act.

Accordingly, Gandhi never tired of praising Tolstoy. He
claimed that Tolstoy's writings are so "good and simple that
a man belonging to any religion can profit by them."21 He
even justified and defended some of the contradictions in
Tolstoy: "the seeming contradictions were no blot on him."
In support he cited Emerson's authority that "foolish consistency is the hobgoblin of little minds."22 We would be utterly
lost, Gandhi proclaimed, if we tried to pursue a life that was
free from contradictions. Observance of a contrived harmony
shall inevitably push us to have resort to untruth. Indeed,
Gandhi felt that the seeming contradictions of Tolstoy were
indicative of his progressiveness! And so he recommended
Tolstoy's Kingdom of God to all his near and dear ones.23 It
is needless to labor here the fact that Gandhi himself laid
great worth on relating precept to practice and invariably
practiced what he seemed to believe in.

John Ruskin was another in whom Gandhi recognized a kindred spirit, and to whom he was drawn strongly. Gandhi's friend, Henry Polak, gave him Ruskin's Unto This Last at the Johannesberg railway station. The book moved Gandhi tremendously: "The book was impossible to lay aside once I had begun it. It gripped me. Johannesberg to Durban was a twentyfour hours' journey. . . . I could not get any sleep that night. I determined to change my life in accordance with the ideals of the book."[24] He felt so strongly for the ideals enunciated in the book that he later translated it into Gujarati under the title Sarvodaya.

Henry David Thoreau is yet another writer from the West to whom Gandhi claims his intellectual indebtedness, even though he does not include him among the prime influences. In reply to a query by an American journalist, Webb Miller, as to whether he had read Thoreau, Gandhi at once said: "Why of course I read Thoreau. I read Walden first in Johannesberg in South Africa in 1906 and his ideas influenced me greatly. I adopted some of them and recommended the study of Thoreau to all my friends. . . . There is no doubt that Thoreau's ideas greatly influenced my movement in India."[25] Similarly, in his letter to Thoreau's biographer, Henry Salt, Gandhi admitted that Thoreau's Essay on Civil Disobedience had "left a deep impression" on him: "The essay seemed to be so convincing and truthful that I felt the need for knowing more of Thoreau and I came across your life of him, his Walden and other shorter essays all of which I read with great pleasure and equal profit."[26]

Among the minor influences on Gandhi were Mazzini and Emerson. What appealed to him in Mazzini's Duties of Man was the latter's teaching that "every man must learn how to rule himself."[27] A similar idea in Socrates also seems to have attracted Gandhi: "Socrates gave us some idea of man's duty."[28] Gandhi read Emerson with considerable interest and advised his cousin also to read his works, for Gandhi found that Emerson's essays contained "the teachings of Indian wisdom in a Western garb."[29]

Ascertaining intellectual indebtedness--especially when a person himself claims or feels to be indebted--is an exercise not without its hazards. And determining what Gandhi, a most complex personality if there ever was one, owed to whom and how much is virtually a hopeless, exasperating task. Scholars often mention the personages above in juxtaposition with Gandhi.

The temptation to compare Gandhi to a wide range of historical personages is great, of course. But such comparisons at best highlight certain values that Gandhi might have cherished in common with other famous men. Otherwise, these comparisons are quite meaningless. He had no chance of interacting with them at a personal level that over time induces the influence that shapes the recipient. The comparisons neither illumine aspects of Gandhi's profound personality, nor do they establish in a very satisfactory manner that Gandhi's ideas and conduct were directly conditioned and coursed by these influences.

Thus, instead of correlating specific influences to particular traits in Gandhi's mental makeup, it is far more rewarding to evaluate the following issues: To what degree was he a product of his milieu? What challenges did his environment pose, and how did he, a creature of complex, conflicting traditions, respond to them? What were the nature and quality of his milieu and his personal temperament? Finally, what was the uniqueness of the vision of life he endeavored to project?

The idea is not to minimize the impact of various personages on him, but to offer instead a somewhat informed corrective to popular generalizations regarding the supposed influences on him. Did Gandhi express his indebtedness to others out of humility? We know how bashful he was inclined to be, especially in his earlier years. Did he accept the thoughts of other celebrities because of their sterling merit, in their own right, so to speak? Or did he take them as a welcome systematization relevant to his situation, and a positive corroboration of his own views and perception of the issues facing him, particularly on the question of war and peace?

Gandhi's Autobiography, considered by some to be "as remarkable as the Confessions of Rousseau," should be an invaluable source for deciphering the development of Gandhi's mind.[30] The work does help in doing this but at the same time hinders analysis. The reader may well be so swayed, if not hypnotized, by what Gandhi himself says as to pay scant attention to the totality of the situation and may thus overlook the particular events and forces that really brought out what lay dormant in him. The extent of influences in that case becomes exaggerated. Gandhi himself was rather prone to do so, also because of his inherent humility, which made him diffident to take the credit for achievements. We, therefore, suggest that not too much importance be accorded to his professed indebtedness lest some of his routine, less glamorous personal experiences—which, in our view, are highly significant—are overlooked or eclipsed.

Without doubt, the most formative period in Gandhi's life was when he lived in South Africa. Until the writing of Hind Swaraj, his life can be said to be a "study in acculturation,"[31] an unfolding story of his reactions and responses to Western culture and civilization. Arriving in South Africa in 1893, he noticed the prevailing racial discrimination there, of course. But the perniciousness of the evil was rudely borne upon him only as a result of personal experiences.

These, in retrospect, Gandhi came to regard as the turning point of his life: the force and intensity of his feelings shifted the compass of his thoughts forever, thereby defining his career with a rare passion and an irrevocable finality.

To represent his employer, Dada Abdullah, Gandhi had to travel to Pretoria. He got into a first-class compartment at Durban with a valid ticket. During the journey the train halted at night at Pietermauritzburg, located in the mountains. A white passenger entered the compartment, looked at Gandhi and turned away to return with a railway official. The official ordered Gandhi to the luggage van in the train. Possessing a valid first-class ticket, Gandhi refused to oblige and was forcibly put down from the train.

Having been thrown out of the train into the bitterly cold night, Mohandas Karamchand Gandhi began to analyze the situation with remarkable calmness and clarity. He had never before suffered such personal indignity. Miserable with cold, as he sat in the dark waiting room of the station, Gandhi debated within himself whether to pocket the insult and proceed on to his destination, Pretoria, or to turn back and pack up for India.

"The cowardly desire to run fought with the determination to stay and fight."[32] Gandhi continues:

> I began to think of my duty. Should I fight
> for my rights or go back to India, or should
> I go on to Pretoria without minding the in-
> sult. . . . It would be cowardice to run back
> to India without fulfilling my obligation. The
> hardship to which I was subjected, was super-
> ficial--only a symptom of the deep disease of
> colour prejudice. I should try, if possible, to
> root out the disease and suffer hardships in
> the process.[33]

His sensitivity prevailed upon him, eventually, and as the dawn streaked the cold hills, Gandhi's mind had been made up: he headed toward Pretoria. As Louis Fischer pithily puts it, "the germ of social protest was born."[34]

Mohandas Karamchand Gandhi's mission in life had begun. Committing himself to the attainment and furtherance of common good, he decided to fight evil. Realization dawned upon him that fear is the greatest evil, be it in an individual or be it in the society as a whole. Justice could be secured by following one's bounden duty to be fearless, by adhering to what one holds to be the truth rather than submitting to what one may feel to be wrong, and by being steadfast and "willingly and peacefully accept[ing] any pain and suffering that steadfastness involves."[35] The lot of his countrymen in

South Africa made abject by the humiliating disabilities they had to suffer constantly and without any hope or sign of redemption touched him intensely and would give him no rest. He resolved to liberate the Indian community. Gandhi's entire stay in South Africa is a saga of his nonviolent struggle for the attainment of justice and dignity for his countrymen there.

At Pietermauritzburg, thus, Gandhi decided to struggle against injustice nonviolently. At that time he had not even seen Tolstoy's Kingdom, for he suffered the humiliation at the station within a couple of weeks of his arrival in South Africa.36 He could not quite reconcile the white South Africans' professions of Christianity with their galling practice of color prejudice. Moreover, as he was currently searching for a social gospel, a gospel of applied religion, he was automatically drawn to Tolstoy's book when he came across it a little later; eventually, Gandhi became overwhelmed by its teachings.

It is, therefore, clear that ahimsa was far more deeply rooted and natural to Gandhi than to have been implanted by his reading of Western sources. Articulation of the idea of ahimsa in these sources pleasantly surprised him. It corroborated his own feelings, and he was thrilled that writers of the eminence of Tolstoy, Ruskin, and others held views similar to his own. At the same time, it convinced him beyond all doubt that ahimsa was moral, opportune, and more than likely to be efficacious.

Thus, when he says that he was "converted" by Tolstoy's book--"I was at that time a believer in violence; its reading cured me of my scepticism and made me a firm believer in Ahimsa"--Gandhi is, in fact, proclaiming confirmation of his own innermost feelings.37 He could hitherto not express such feelings for a variety of reasons including his bashful nature, modesty, and an almost total lack of reading.38 His enthusiastically loud eulogy of the Western writers is by no means the reaction of a mind already enriched by extensive reading and deep deliberation.

His lack of reading at that time--though later, of course, he did read a lot--explains quite a few things. The reason why the relatively minor works of Tolstoy, Ruskin, and others acquired great significance for Gandhi was that he read not systematically but haphazardly and what came his way accidently. His intellectual hesitancy and lack of confidence were so manifest and pervasive that even on ordinary, commonplace issues Gandhi would quote extensively from these writers, using them as backstage prophets. For instance, on the ill effects of tobacco, he would seek authority in Tolstoy: "Tolstoy called it the worst of all intoxicants;" on his impressions of Paris, particularly the Eiffel Tower, he would be as deprecating as Tolstoy had been: "I remember that Tolstoy was the chief among those who disparaged [the tower]. He said that the Eiffel Tower was a monument of man's folly not of his wisdom;"[39] on being asked at Lausanne, returning home in December 1931 from the London Round Table Conference, as to what his message for the women of Europe was, he simply reproduced Tolstoy's view of women: "As Tolstoy would say they are labouring under the hypnotic influence of men;"[40] and so on.

And yet too much cannot be made of Tolstoy's influence on Gandhi even as it is reflected in the correspondence, during 1909-10, between the two. The correspondence has to be read, as we suggested earlier, against the background of the actual circumstances in which it was written.

Gandhi first wrote to Tolstoy on October 1, 1909,[41] from London, seeking permission to print Tolstoy's Letter to a Hindu, which was a monograph of sorts written by Tolstoy in response to some challenging questions from Indian revolutionary editors of an underground journal, Free Hindustan, issued from Vancouver.[42] In the monograph, Tolstoy, rejecting the argument of the Indian revolutionaries, asserted categorically, "Love is the only way to rescue humanity from all ills, and in it you too have the only method of saving your people from enslavement."[43] Such a sentiment was very much after Gandhi's heart, for it buttressed his technique of satyagraha; hence his request to Tolstoy. Soon thereafter he sent Tolstoy a copy of Hind Swaraj. In the accompanying letter he described himself a "humble follower" of Tolstoy.[44]

It would be wrong, however, to interpret this courtesy as an indication that Gandhi's subsequent life was guided by the pacifist writings of Tolstoy, for a passionate belief in the philosophy of nonviolence, duly reflected in Hind Swaraj, was already the bedrock of Gandhi's thinking and the guiding light of the course of his action. This is what must have created fondness between them and established the enduring bond of mutual respect. Tolstoy replied, "I read your book with great interest because I think that the question you treat in it--the passive resistance--is a question of the greatest importance not only for India but the whole humanity." On Tolstoy's death in November 1910, Gandhi wrote in homage: "For inculcating this true and higher Ahimsa amongst us, Tolstoy's life with its ocean like love should serve as a beacon light, never failing source of inspiration."[45]

Thus, Tolstoy's message of universal love had a great appeal for Gandhi, but was it not already there at the Pietermauritzburg railway station in 1893 when he decided to wage a struggle for his countrymen? Of course, Gandhi confessed to be "a devoted admirer" of Tolstoy.[46] The word "admirer" is significant here, not only because it denotes

Gandhi's innate humility and personal esteem but also reciprocity to the Russian's declared admiration for Gandhi and his work. In his letter of September 7, 1910, Tolstoy wrote to Gandhi:

> Your activity in Transvaal as it seems to us
> at this end of the world is the most essential
> work, the most important of all works now
> being done in the world, wherein not only the
> nations of the Christians but of all the world
> will unavoidably take part.[47]

The suggestion that Gandhi was merely putting into effect what the noble Russian had preached through his writings is not entirely tenable.[48] We would much rather see Tolstoy's letter as an expression of his satisfaction at the very welcome growth of nonviolent resistance to injustice in the state. The letter reflects not only a vindication of all that he had stood for, but it is also a manifest expression of the hope of a dying man. Gandhi seems to lend substantiation to our interpretation when he says, "I am glad to recall the fact that in a long letter he [Tolstoy] wrote to me unsolicited, he said that his eyes were fixed upon me wherever I was."[49]

That Gandhi was discovering erudite and welcome confirmation of his own convictions in these celebrities is further borne out in his article, "To American Friends," dated August 3, 1942.[50] "Russia gave me in Tolstoy a teacher," yet a teacher who does not seem to have taught Gandhi anything new, but to have only furnished a "reasoned basis for my non-violence." As late as just a couple of months before his assassination, Gandhi felt that Harmony of Gospels and Tolstoy's other books helped him to clarify his own ideas on passive resistance.[51]

Similarly, there is no need also to be taken in by Gandhi's statement that Ruskin's Unto This Last made him change the pattern of his life, for he also says that "some of my deepest conviction [is] reflected in this great book . . . and that is why it so captured me and made me transform my life."[52] As final evidence in this regard, we offer the use that Gandhi made of Thoreau's writings. He was already soaked in the spirit of resistance to evil and his satyagraha movement was well under way when Gandhi read Thoreau's essay The Duty of Civil Disobedience. Thoreau, thus, provided Gandhi not with the initial inspiration but merely with a confirmatory rationale. "To American Friends" says, "you have given me a teacher in Thoreau who furnished me through his essay [with] scientific confirmation of what I was doing in South Africa."[53] Gandhi, at the most, borrowed just the term, rather than the idea, of civil disobedience from Thoreau.[54] It is not surprising that after a time he dropped Thoreau from his list of those who had influenced him.[55]

In brief conclusion, then, we are inclined to agree with his biographers, Polak, Brailsford, and Pethick-Lawrence, that Gandhi "was confirmed in this view by his reading of Tolstoy's writings and of . . . Thoreau's essay on the Duty of Civil Disobedience."[56] It would certainly be very wrong to take a mechanistic view of the extent and intensity of others' influence on a mind that is great in its own right. Great minds absorb congenial influences: like welcome rain helps the seeds already there to germinate and flourish. Gandhi was touched by these books, which said nothing to millions, simply because he already had the seeds of conviction within him. Concerning his vegetarianism, he remarks:

> I was born a vegetarian. I was a vegetarian
> by the vow I made before my mother. Then
> I read Salt's Plea For Vegetarianism and I was
> confinced.
>
> But the conviction was already in me. Sim-
> ilarly, with Ruskin's Unto This Last. I was
> trying to follow that life, but Ruskin made it
> real in my own life. He changed it, but the
> conviction was already there. To others, in
> whom the conviction was not already there, the
> same book would make no appeal."[57]

It is clear, therefore, that conviction already existed in Gandhi; all he sought and got by exposure and receptivity to these influences was the weight and "respectability" from personages of established eminence.

The origin of Gandhi's thinking on war seems to be shrouded in a haze of inarticulation, or eclipsed or encompassed by the umbrella of his general philosophy. There is, however, a curious little episode dating back to his student days in London, which may well be considered to have brought Gandhi to comprehend what, in terms of human suffering and degeneration, war means.

Torn by doubts and anxieties about his legal career, Gandhi was advised to meet Frederick Pincott, a Conservative with genuine affection for Indian students. Gandhi said that it was a memorable meeting: "I can never forget that interview." Pincott advised him to read "to understand human nature," and among the books he suggested was Kaye and Malleson's History of the Mutiny of 1857. Gandhi could not read it in England but did so in South Africa, as he "had made a point of reading [the volumes of Mutiny] at the first opportunity."[58] Some think that there is no way of knowing the impression this work made on Gandhi.[59] But, in fact, he seemed to have quite vivid recollections of it. In

1931 Gandhi urged Congress <u>Seva Dal</u> workers in Bombay to
read history with his eyes, especially the history of the
mutiny, and held that the mutiny

> was a war of independence fought with violent
> weapons. Colonel Malleson has narrated a fair-
> ly faithful account. You will see that though
> the greased cartridges may have been an im-
> mediate cause, it was just a spark in a maga-
> zine that was ready. But look at the result.
>
> The U.P., the storm centre of 1857, has for
> generations since remained under a paralysis as
> perhaps no other province. For people have re-
> tained vivid memories of man turned beast and
> masses who simply watched were mown down like
> corn stalks in a field.[60]

Again, his observations on Panjab massacres in the wake
of the partition of the country in 1947 show him drawing
parallels and lessons from the mutiny. Referring to the
method and use of violence, thus, he felt that some sort of
peace might perhaps be clamped down on bloodbathed, burning Panjab by means of brute force, but it was likely to be
precarious and debilitating. He sincerely hoped--but was not
certain--that the poison engulfing Panjab would not spread
to the rest of the country, as it had in 1857:

> Similar things . . . had happened during the
> Sepoy War, when it was quelled by means of
> superior arms.
>
> Outwardly, things quietened down but the
> hatred against an imposed rule went deep
> underground, with the result that we are even
> today reaping the harvest of what was then
> sown. The British Government took the place
> of the East India Company . . . and imposed
> political subjugation and inflicted degradation
> on India.[61]

Thus, the imprint of Kay and Malleson's book on Gandhi
apparently is unmistakable and deep. Indeed, it reinforces
his steadfast faith in complete rejection of violence, however
warranted by the circumstances it may seem to be.

Harsh realities of actual war replacing the remoteness of
a hazy historical phenomenon, however, forced themselves
on Gandhi soon after he had read Kay and Malleson in South
Africa. He saw at first hand, as a stretcher bearer in the
front lines of the Boer War (1899-1902), the utter havoc
warfare wreaks on the bodies and minds of those involved.
Although Polak and associates felt that his actual experience

of the Boer War fully matured Gandhi's views, in reality his
reactions at the time were rather mixed: he was deeply
struck by the heroism, and the readiness to die, shown by
the brave soldiers in that war; the battlefield strangely recalled to his mind the peace and activity of the Trappist
monastery in Natal. Indeed, in a speech in early October
1902, he actually compared the holy stillness pervading a
Trappist monastery he had visited to the atmosphere he had
experienced in some war camps, with their perfect stillness
and dedicated preparedness to put forth the maximum energy
and selflessness in the face of a great ordeal:

> Tommy [British soldier] was then altogether
> lovable. . . . There was, shall I say, a
> spirit of brotherhood irrespective of colour or
> creed.
>   As a Hindu I do not believe in war, but if
> anything can even partially reconcile me to it,
> it was the rich experience we gained at the
> front.
>   It was certainly not the thirst for blood
> that took thousands of men to the battlefield.
> If I may use a most holy name without doing
> any violence to our feelings, like Arjun, they
> went to the battlefield, because it was their
> duty. And how many proud, rude, savage
> spirits has it not broken into gentle creatures
> of God?62

He was so dazzled by the apparent dedication of the
soldiers to duty and discipline--values he cherished so
immensely--that the ugliness of the war perhaps failed to
register. Clearly, at this period in his life, Gandhi did not
deplore war as he was to later on. We shall have occasion
to examine his experiences of the time at some length in the
following chapter.
      Death danced again in the sight of Gandhi during the
Zulu Rebellion (1906), where once more he tended lacerated
bodies and bewildered souls of the hapless Zulus. Here was
a proud people on whom war had been declared by a mighty
government simply because the Zulus refused to pay arbitrary taxes. It was probably this bloody experience that
turned Gandhi away from the method of war irrevocably:

> The Zulu "rebellion" was full of new experi-
> ences and gave me much food for thought.
> The Boer War had not brought home to me the
> horrors of war with anything like the vividness
> that the "rebellion" did. This was no war but
> a man-hunt. To hear every morning reports of

soldiers' rifles exploding like crackers in
innocent hamlets and to live in the midst of
them was a trial.[63]

The experience, naturally, set Gandhi soul searching and
scrutinizing the phenomenon of war. The conclusion he arrived at was anticipated. Speaking at the Emerson Club in
London on October 8, 1909, he expressed himself for the
first time strongly in condemnation of war:

War with all its glorification of brute force
is essentially a degrading thing. It demoralizes
those who are trained for it. It brutalizes men
of naturally gentle character. It outrages every
beautiful canon of morality. Its path of glory
is foul with passion and lust, and red with
blood of murder. This is not the pathway to
our goal.[64]

Apparently then, the realization of war being an unmitigated evil, utterly indefensible under any circumstances,
came to Gandhi not from his reading of the likeminded celebrities, but out of compassion aroused by intimate and firsthand witnessing of the cruelties of war.

Tolstoy too had rejected war after his personal experience of it in the Crimean War. In command of a battery at
Sebastapol, the sight of bloodshed and carnage made him
"brim over with compassion." The world became richer by
his classic War and Peace which exposes how war leads to
total moral degradation. Henceforth, Tolstoy never ceased
to strip militarism of its glory nor overlook an opportunity
to expose its brazen falsity and brutality; all its fine titles
and trappings could scarcely conceal the true purpose of
war: murder, nothing more nor less. And so, Tolstoy, in
utter revulsion of war, tried to evolve the technique of passive resistance from out of the old theological doctrine of
nonresistance. All he, eventually, came up with was nothing
more than the suggestion that something more than nonresistance could lead to the elimination of war. As to how
peace was to be created, Tolstoy had precious little to suggest.

Gandhi's concern, in contrast, was much more than mere
elimination of war. He was primarily concerned with evolving
a social order that while ensuring peace would automatically
preclude all possibilities of war. Mention may be made en
passant that Gandhi probably never read Tolstoy's classic.
But for a casual, cryptic remark that Tolstoy "wrote a very
effective book on the evil consequences of war," neither
Gandhi nor any of his biographers includes War and Peace
among Tolstoy's works read by Gandhi.

In fact, Gandhi could just not reconcile wars, especially those of Europe, with Christianity. "It is a very curious commentary on the West," he told journalist Andrew Freeman of the New York Post, "that although it professes christianity, there is no christianity or Christ in the West, or there should have been no war. That is how I understand the message of Jesus."65 Gandhi had, apparently, found his own thinking on the subject echoed in Tolstoy's Kingdom of God according to which "there is an irreconcilable antagonism between the christian faith and war." Indeed, he maintained that "Europe does not believe in [the New Testament]. . . . They do claim to respect it, although only a few know and observe Christ's religion of peace."66

Here, evidently, is the reason as to why the Old Testament, replete with accounts of violence and wars, had no appeal for Gandhi; in fact, he found it repellent. Even while he admires the Boers for their courage and rugged faith, he strongly disapproves of their violence which he feels rested on too literal an interpretation of the Old Testament. The Boers, he held, "read the Old Testament with devotion and knew by heart the descriptions of battles it contains. They fully accept Moses' doctrine of 'an eye for an eye and a tooth for a tooth' and they act accordingly."67 He supposed that the Boers knew the New Testament only by name! Gandhi held Christians squarely responsible for present-day wars, which put to shame even those described in the Old Testament.68 The reason why there were wars in Europe, Gandhi strongly felt, was that the Christian world was not following the message of Christ or the creed of the New Testament, for there simply was no justification for war in that text.

Yet all this should not blind us to the basic fact that Gandhi's view of war comes, decisively, from his traditional Hindu moorings. Stephen Hay has argued (see footnote 15) that there is a great deal in Gandhi's "way of acting and thinking that makes more sense in the context of Jaina goals and disciplines than of Hindu ones." This, however, is certainly not the case in regard to Gandhi's views on war whereupon his ideas are overwhelmingly disciplined by Bhagwad Gita, a Hindu text. Without doubt, Gandhi interpreted war, and participation in it, strictly in the light of the Gita. Indeed, it is quite apparent that on the questions of war and peace Gandhi set himself the task of reconciling the Gita with the Sermon on the Mount, to work the sermon into the framework of the Gita, even at the expense of considerable disapproval.69 Be it noted en passant that, chronologically, Gandhi's tentative comments on the subject of war appeared toward the conclusion of the Boer War in 1902 rather than as a reaction to the Russo-Japanese War (1904-05) as some students of Gandhi erroneously hold.70

The Gita, held Gandhi, deals with a person's duty in time of crisis. It is not a historical discourse; rather, it merely uses physical illustration to drive home a spiritual truth.71 Just as he treated the Son, the Father, and the Holy Ghost of the New Testament merely figuratively, similarly he refused to belittle the teaching of the Gita "by confining it to a description of a clash of arms between rival clans on an earthly battlefield." He maintained that the description in the Gita was "not of a war between cousins but between the two natures in us, the good and the evil;" though good is always victorious, evil also does put up a brave show and baffles even the keenest conscience. He said the battle described therein conveys, primarily, the priceless teachings. The author of the Gita, poet Ved Vyasa, did not look upon war as something morally wrong, for in his times, war was not a taboo, nor was it "considered inconsistent with ahimsa." Thus, to argue that the Gita teaches violence, or justifies war, was in Gandhi's opinion as "unwarranted as to argue that since violence in some form or the other is inescapable for maintaining body in existence, dharma lies only in violence." Convinced that the Gita neither advocates war, nor condones violence, he regretted that terrorists, in their perfect but misguided honesty and earnestness, used the book to defend their policy and methods. Merely dismissing his interpretation and holding on to their own carried no weight, for only time will show whose reading of the text was correct and relevant.

Whether Gandhi's interpretation was right or not, the fact is that his view of the Gita and the conclusions he drew regarding its significance are, indeed, crucial to the understanding of the development and shaping of his ideas on war and peace.

Commenting upon the Statesman's censure of the "War against War" movement in Britain led by a British conscientious objector, Canon Sheppard, in the mid-1930s, Gandhi said that the dragging in of the Gita by the newspaper in support of its criticism was quite misplaced. The position of the conscientious objectors in the West, he pointed out, was rather different from that of Arjuna, who actually believed in war: the dilemma he faced was not one of love of men versus hatred of war, but simply that how could he brace himself to kill his own kith and kin. Under the circumstances, the only advice that Krishna could give was the one he actually gave: "do your duty;" or in other words, fight the evil.

Gandhi was convinced beyond doubt that Vyasa "wrote his supremely beautiful epic to depict the futility of war;" the manner in which the omnipotent Krishna, the invincible Arjuna, and the Dharmaraja Yudhishthara were eventually decimated, weakened to impotence, or condemned to perdition

clearly supports this. Thus, the Gita proved to Gandhi the
"utter nothingness of earthly power." Having arrived at this
view in the mid-1920s, Gandhi seems to have stuck to it till
his very end. Writing in September 1936, he reiterated that
the author of the Gita had shown the world "the futility of
war by giving its victors an empty glory." This attitude is
vividly reflected in Gandhi's conjecture regarding World War
II, expressed in 1942:

> Personally I think the end of this giant war
> will be what happened in the fabled Mahabharata
> war. . . . What is described in that great epic
> is happening today before our very eyes. The
> warring nations are destroying themselves with
> such fury and ferocity that the end will be
> mutual exhaustion. The victor will share the
> fate that awaited the surviving Pandavas.[72]

More than revealing war as a delusion and folly, the
Mahabharata to Gandhi was "a sermon on non-violence. Fight
without anger and passion can only be spiritual." The Gita
laid great stress on the significance of truth, and the
"Mahabharata sums up its teaching by declaring emphatically
that truth outweighs everything else on earth." By bringing
out the futility of war, and establishing an inexhorable relationship between the ends and means, the Gita "shows the
only way to right action."

On being asked how a righteous battle, as depicted in
the Gita, could produce catastrophic results, Gandhi explained it was "because of the means used;" for noble results the means employed should equally be noble; means
should not be distinguished from ends; if violent means are
used the result is bound to be bad. This, asserted Gandhi,
was true and valid for all times. With regard to World War
II, he sadly noted that the laudable end of defeating fascism
had been achieved by decidedly bad means by England and
America: "this was not the way of truth."[73]

In sharp contrast to the Marxist position that every war
should be judged by the historical ends it serves, and that
certain wars are justified insofar as they destroy extremely
pernicious and reactionary institutions, Gandhi is firmly of
the opinion that war is not a morally legitimate means of
achieving anything permanent: enduring peace through war,
or justice through unjust and dubious means, according to
Gandhi, was simply not feasible.

Beside the futility of war, what else did the Gita signify
for Gandhi? His ultimate, mature conclusion about the book
was that instead of the rules of warfare it reveals the qualities of a perfect human being. The reason is that it calls
upon the individual not to cooperate with the forces of evil

and darkness. At the same time, it urges that one should go on working without expecting, longing for, or being attached to the fruits of one's work. To use Gandhi's own words, "I deduce the principle of satyagraha [nonviolent resistance] from this. He who is free from such attachment will not kill the enemy but rather sacrifice himself." The urge to kill an enemy, Gandhi maintained, proceeded from impatience, which, in turn, arises from attachment. He saw justification of his view--namely, "sacrifice yourself [rather] than spill the blood of others"--in the stand President Kruger took in the Boer War. Kruger had said that he would "stagger humanity," meaning that he would sacrifice every Boer--man, woman, and child--rather than leave a single Boer to be slaughtered; he would gladly let the English roam about the ravaged and deserted soil of South Africa "dyed with the blood of Boer martyrs."[74] And England, Gandhi pointed out, had to yield since it was tired of concentration camps in which Boer women and children died like flies and was choked with the bloody feast that the dead Boers provided it.

The activist in Gandhi, therefore, sees that a person has to act, against violence and evil--that is the lesson of the Gita for Gandhi. At the same time, Gandhi is quite firm that the Gita dharma does not mean that a person who has not so far awakened to the truth of nonviolence may act like a coward. This dharma does not countenance running away in fear under any circumstances. It is in this context that violence is not ruled out by the Gita: "I do not wish to suggest," says Gandhi, "that violence has no place at all in the teaching of the Gita."[75]

In this world of ours which baffles all reason, violence will, then, always be. The Gita shows the way out of it, of course, but it is not the way of cowardice; people cannot escape violence simply by running away like cowards. Indeed, Krishna teaches that one who tries to slink away would do better, instead, to kill and be killed: a gospel of duty in the face of evil. Our mission in life is to take action against evil rather than to avoid activity. Hence, Gandhi chose to act, rather than remain idle or aloof, during certain wars as we shall see later.

Such an attitude contrasts sharply with what Tolstoy advises: the individual should abstain from participation and refuse conscription for the reason that the standing armies invite war. An increase of armed forces not only fails to protect us from dangers of attacks by others, holds Tolstoy, but actually provokes and precipitates the very attack it deprecates.[76] Gandhi, on the other hand, did not preach against conscription. In fact, he would allow it wherever it was necessary: "If the whole country . . . felt the necessity of military training, why should there be any objection to giving it a place in the law of the land?"[77]

In his book <u>Unto This Last</u>, Ruskin held:

> It is one very awful form of the operation of
> wealth in Europe that it is entirely Capitalists'
> wealth which supports unjust wars. Just wars
> do not need so much money to support them,
> for most of the men who wage such wars, wage
> them gratis; but for an unjust war, men's
> bodies and souls have both to be bought; and
> the best tools of war for them besides; not to
> speak of base fear, and angry suspicion, which
> have not grace nor honesty to buy an hour's
> peace of mind with. . . .
> And all unjust war being unsupportable, if
> not by pillage of the enemy, only by loans
> from capitalists, these loans are repaid by sub-
> sequent taxation of the people who appear to
> have no will in the matter, the Capitalists will
> bring the primary root of war.[78]

Further, in his work <u>The Crown of Wild Olive</u>, Ruskin also
supported the concept of "just" war: "You soon will have no
more war unless it be indeed such as is willed by HIM, of
whom, though Prince of Peace, it is also written in righteous-
ness He doth judge and make war."[79] Such literal reading of
the Bible distressed Gandhi as much as he was at the Boers'
literal subservience to the Old Testament.

Gandhi simply did not accept the distinction between
"just" and "unjust" wars; for him every war was wrong.[80]
Each war entailed immense violence, and under the label of
"just" war, use of violence was sought to be justified on
moral and social grounds. This, in Gandhi's view, was quite
indefensible. He did side with those whose cause was just on
ideological grounds; for instance, he was sincerely on the
side of the Western democracies in combat with Fascists and
Nazis in World War II. But he unequivocally and emphatically
disapproved of the violent means employed in the so-called
just war, for the simple reason that he passionately believed
in the superiority and sublimity of nonviolent means, whatever the situation.

Gandhi rejected the distinction between "just" and "unjust" war, it is sometimes assumed, under inspiration from
Thoreau. That is quite wrong, however. Even after he had
read Thoreau's <u>Civil Disobedience</u>, he cooperated with the
British in India to raise recruits for World War I. Thoreau's
essay contains his reaction to the Mexican War of 1848: he
had refused to pay the poll tax and advised others also to do
the same, because he felt that the money so collected was
going to a wrong cause.[81] Gandhi, on the other hand, did
not believe in boycotts of this type. As we shall see, Gandhi

was, in fact, much more than a nonpayer of taxes. In a system based on violence, and one to which he must belong without the slightest option, Gandhi would much rather participate directly than remain inactive or indifferent:

> If I have only a choice between paying for the army of soldiers to kill my neighbours or to be a soldier myself, I would, as I must consistently with my creed enlist as a soldier in the hope of controlling the forces of violence and even of converting my comrades."82

The entire phenomenon of war was repugnant to Gandhi. But he had no use at all for the pacifism of Quakers or conscientious objectors or war resisters as an ideology of protest on grounds merely of conscience. Contrary to the pacifist position, his indeed was the stance that in a conflict there should always be ardent and constant participation rather than withdrawal or apathy. Pacifism, as a moral attitude, did not have any appeal for him: "To refuse to render military service when the particular individual's time comes is to do the thing after all the time for combating evil is practically gone. . . . Refusal of military service is much more superficial than non-cooperation with the whole system which supports the state."83

In fact, unlike the pacifists, Gandhi's attitude toward war was not as if a given cult were to be adopted; rather it was a natural outcome of his own perception and experience. His problem was quite different from that of the pacifists. He was not concerned with war per se. Whereas the starting point of the pacifists was "negation of war," Gandhi was, primarily and very intensely, drawn to the question of human dignity in all its aspects. It was in this context, and with this fundamental consideration, that he encountered and reacted to war. His views on war and peace, accordingly, arose and evolved in the series of his experiments with truth based on nonviolence, total identity of means and ends, self-suffering, and endeavors to convert the opponent.

The firm and unequivocal fundamental for Gandhi was action, which is inexorably and inextricably linked with reality. That is why the Gita had such an indelible and deep impact on him.84 Clearly, he took to heart the injunction of the text: "He who gives up action falls." It is precisely for this reason that in time he dropped Emerson and Mazzini from his list as inactive intellectuals. The bania in him enabled him quickly to differentiate essentials from nonessentials!

The intellectual gymnastics of philosophy and metaphysics never held any fascination for Gandhi. He was always for definite, exact action. It is this powerful, overwhelming instinct for action that draws Gandhi close to the Gita, for the text proclaims positive action based on nonattachment, release from fear, hatred, pride, and anger, thereby pointing to the path of selfless action. This unflinching faith in the ideal of action under dictum of duty drawn from the Gita, naturally then, determined Gandhi's attitude to war: "When two nations are fighting the duty of a votary of ahimsa is to stop the war. He who is not equal to that duty, he who has no power of resisting war, he who is not qualified to resist war, may take part in war, yet whole-heartedly try to free himself, his nation and the world from war."[85] We shall note in the next chapter how Gandhi explained his peculiar participation in war in terms of his duty as delineated by the Gita: "What was my duty as a citizen of the empire as I then believed myself to be . . . I therefore felt that I sufficiently discharged my duty as a man and a citizen by offering my humble services to the empire in the hour of its need."[86]

Vincent Sheean rightly maintains that after the Zulu Rebellion, the 37-year-old Gandhi based the conduct of his entire subsequent life upon the Bhagwad Gita.[87] Gandhi had held the Gita supreme from the very earliest days of his reading it, of course, but his yen for it grew with the years. His observations to take the physical combat at Kurukshetra allegorically notwithstanding, there is little doubt that Gandhi took all its great concepts and teachings literally, except, of course, the bloody framework in which they are set. He disclaimed all intention of straining the meaning of the book to suit any preconceived notions or motives of his; indeed, "my motives were the outcome of a study of Gita."[88]

An extremely important principle that the book seems to have taught him, which actually determined his entire behavior and conduct throughout life, was apalayanama, that is, refusal to run away, withdraw. In the epic Krishna asks of Arjuna--or anyone faced with evil, for that matter--precisely this. Arjuna was not unwilling to fight but developed qualms of conscience and remorse only over fighting his own kinsmen. Gandhi read in it an attempt on the part of Arjuna to run away from the situation, not doing his duty; his inaction would invite dishonor, so to Gandhi the only way out for Arjuna was "action" based on detachment, to perform the duty expected of one's post and station in life. To use the words of his admirer, George Catlin, Gandhi interpreted the Gita as a gospel of duty and action against the evil passions.[89]

In the Gita Gandhi also found the sanction for ahimsa as the supreme law. According to him Gita teaches anasakti, that is, the path of selfless action, and anasakti transcends ahimsa. He who would be anasakta, that is, selfless, has of

necessity to practice nonviolence in order to attain the state
of selflessness. "Ahimsa is, therefore, a necessary preliminary, it is included in anasakti, it does not go beyond it."90

Gandhi had neither the intention, nor indeed the pre-
tension, to be a theorist. He was essentially and totally a
man of action, an actionist so to speak: "I am not built for
academic writings. Action is my domain."91 The means of
acquiring knowledge, for him, was action and not mere contemplation. Rather than sit back and contemplate, Gandhi
was up and about to change the existing order of things.
Moving from truth to truth, he made introspection as the
sole guide to his action. In an unqualified devotion to the
ideal of "duty and action," Gandhi lays stress on self-
suffering, maintaining that the logical outcome of "the
teachings of the Gita is decidedly for peace at the price of
life itself." He admired Jesus, for the latter "defended his
honour and that of man when he preferred the death of a
felon to the denial of his faith." To Gandhi, death is more
honorable than the destruction of faith and meek surrender.
He emphasized that the Bible teaches the law of love, which
is the very law of our being; were it not so, life would not
have persisted in the midst of death. Modern science renders possible the impossible, but "the victories of physical
science would be nothing against the victory of Science of
Life which is summed up in Love which is the Law of our
Being."92

Gandhi attributed the recurrence of war and the absence
of permanent peace to the want of a living faith in a living
God. He thought it to be a first-class human tragedy that
people of the earth, who claim to believe in the message of
Jesus whom they describe as the Prince of Peace, show little
of that belief in actual practice. How painful it is to see
sincere Christian divines limiting the scope of Jesus' message
to select individuals!

Indeed, his boundless admiration for the New Testament
version of the Christian message encouraged Gandhi to hope
and urge the key feature of his blueprint for a durable
peace, namely, a unilateral disarmament, as we shall discuss
in Chapter 7: "If even one great nation were unconditionally
to perform the supreme act of renunciation, many of us
would see in our lifetime visible peace would be established
on earth"[Gandhi's italics].93 Peace to a war-prone and
often war-torn world, in Gandhi's view, could come only
through selfless action and compassion. His approach to the
problem of peace, it seems clear, was shaped by his understanding of the Gita and the Sermon on the Mount. Both
these sources inspired him, and Gandhi derived great com-
fort and encouragement from the New Testament--which has
"nothing but love in every page"--since that was exactly
what he had in his mind. He saw no "difference between

Sermon on the Mount and the Bhagwad Gita."[94] His innova-
tive genius lay in relating the Gita's exhortation to duty to
the mellow compassion of the sermon for seeking the truth.

It is thus obvious that Gandhi was not merely a recep-
tacle or resonance of the various influences to which he was
exposed in his life, particularly those of certain individuals.
This fact is sharply noticeable in regard to his attitude and
approach to the whole question of war and peace. Some,
especially among Quakers, recognized it; in the words of
pacifist Reginald Reynolds who came in personal contact of
Gandhi, "A false antithesis having been made between peace
and justice, christian pacifists in most cases chose a passive
conception of peace . . . It is from this ethical fog that
Gandhi has done much at last to rescue us."[95]

Tolstoy, Ruskin, and Thoreau's thinking on the subject
hardly touched Gandhi, except perhaps in a very general
way as a possible stimulus. He had steadily evolved a peculiar synthesis of his own on the issue. His technique for
combating war, termed by Erik Erikson as the philosophy of
militant nonviolence, emerged probably out of the event that
occurred in Ahmedabad in 1918.[96] It became an effective
political instrument transcending the immediate context of,
to use Erik Erikson's expression, industrial peace, meaning
the peaceful resolution of a labor-management dispute in the
textile industry there. The approach and the instrument so
forged seemed pertinent to the world of the day. Gandhi
believed that it promised the great potential of effective application on as large a scale as that of the whole globe. The
synthesis transformed him into the Gita's definition of karmayogi, that is, a spiritual aspirant devoted to the path of
action. It was in the spirit of the karmayogi that Gandhi
tackled the issues of war and peace among the rest that constituted his universe, and duly won the accolade of mahatma,
a great soul.

NOTES

1. Harijan, vol. 4 (New York: Garland, 1973), p. 49.

2. The Collected Works of Mahatma Gandhi, vol. 10 (New
Delhi: Government of India, Publications Division), p. 189.

3. Mahatma Gandhi, Hind Swaraj (Ahmedabad: Navajivan,
1962), p. 105.

4. Mahatma Gandhi, An Autobiography, My Experiments
with Truth (Ahmedabad: Navajivan, 1959), p. 4.

5. Ibid., p. 20.

6. Vincent Sheean, Lead Kindly Light (London: Cassell, 1950), p. 70.

7. Autobiography, p. 35.

8. Possibly indicative of his modest intellectual tendencies is the fact that in India, before he had gone to England, he "had never read a newspaper." Ibid.

9. See Erik Erikson's psychobiography, Gandhi's Truth, On the Origins of Militant Nonviolence (London: Faber & Faber, 1970), p. 152.

10. Autobiography, p. 65.

11. Ibid., p. 64.

12. The questions are to be found in Pyarelal, Mahatma Gandhi, The Early Phase, vol. 1 (Ahmedabad: Navajivan, 1965), p. 327; Collected Works, vol. 1, pp. 90-91.

13. Collected Works, vol. 37, pp. 261, 265.

14. Autobiography, p. 99.

15. "Her determined resistance to my will on the one hand and her quiet submission to my stupidity involved on the other, ultimately made me ashamed of myself. . . . In the end she became my teacher in non-violence. And what I did in South Africa was but an extension of the rule of Satyagraha she practised in her own life." Harijan, vol. 6, p. 394. Moreover, ahimsa inevitably and easily came to him also from a host of other native sources. Stephen Hay, in "Jaina Goals and Disciplines in Gandhi's Pursuit of Swaraj," has convincingly argued that Gandhi could not possibly help imbibing Jainism--of which ahimsa forms one of the five fundamental disciplines--in degrees from, one, his mother, who had been trained in Jain ways before her marriage and continued to value and be guided by Jain monks all her life (it was a Jain monk, Becharji Swami, who persuaded her to allow Gandhi to go to England for education), two, the Jain monks who visited his father frequently, and three, Ray-chandbhai, of course. Peter Robb and David Taylor, eds., Rule, Protest and Identity (London: Curzon Press, 1978), pp. 120-132.

16. Collected Works, vol. 37, pp. 261-263.

17. Ibid., vol. 9, p. 243.

18. Ibid., vol. 37, p. 262.

19. "It is not that others have not said what Tolstoy said. . . ." Ibid., p. 266.

20. Ibid., p. 267.

21. Ibid., vol. 9, p. 241.

22. Ibid., vol. 37, p. 264.

23. For instance, Gandhi wrote on March 25, 1909 to his son Manilal: "Please tell Maganlalbhai that . . . he should also try to read Tolstoy's Kingdom of God Is Within You. It is a most logical book. . . . What is more, Tolstoy practices what he preaches." Ibid., vol. 9, p. 209.

24. Ibid., vol. 49, p. 241; also, Autobiography, p. 220.

25. Cited in George Hendric, "Influence of Thoreau and Emerson on Gandhi's Satyagraha," Gandhi Marg, vol. 3 (New Delhi: Gandhi Peace Foundation, July 3, 1959), p. 166.

26. Collected Works, vol. 41, p. 553.

27. Hind Swaraj, p. 67.

28. Collected Works, vol. 8, p. 241.

29. Ibid., vol. 9, p. 209.

30. G. Bolton, The Tragedy of Gandhi (London: Allen & Unwin, 1934), p. 2.

31. Chandran D.S. Devanesan, Making of the Mahatma (Delhi: Orient Longmans, 1969), p. 399.

32. Ibid., p. 233.

33. Autobiography, p. 81.

34. Louis Fischer, The Life of Mahatma Gandhi (Stuttgart: Tauchnitz, 1953), p. 37.

35. K.S. Murty and A.C. Bouquet, Studies in Problems of Peace (Bombay: Asia, 1960), pp. 100-101.

36. Gandhi's first biographer, Reverend Joseph J. Doke, informs us that it was during 1893-94 that Gandhi read "quite 80 books which included some of Tolstoy's works."

See his M. K. Gandhi, An Indian Patriot in South Africa (Varanasi: Akhil Bharat Sarv Seva Sangh Prakashan, 1959), p. 59. Also see "An Appeal to Every Briton in South Africa," in Collected Works, vol. 1, pp. 256-286.

37. D.G. Tendulkar, Mahatma, vol. 1 (New Delhi: Government of India, Publications Division, 1960), p. 39.

38. "During the days of my education I had read practically nothing outside text books and after I launched into active life I had very little time for reading. I cannot therefore claim much book knowledge." Autobiography, p. 220.

39. Ibid., pp. 55-56.

40. Collected Works, vol. 48, p. 407.

41. According to Kalidas Nag, Gandhi had sent felicitations on Tolstoy's birthday in September 1908. See his Tolstoy and Gandhi (Patna: Pustak Bhandar, 1950), p. 57. For Gandhi's letter, see Collected Works, vol. 9, pp. 445-446.

42. Nag erroneously identifies the chief editor of the journal as C. R. Das (Tolstoy and Gandhi, p. 76), when in fact the name was Tarak Nath Das.

43. Ibid., p. 91.

44. Collected Works, vol. 10, pp. 210, 505.

45. "The Late Lamented Tolstoy the Great," in ibid., p. 369.

46. Ibid., vol. 21, p. 354.

47. Ibid., vol. 10, pp. 512-513.

48. J.L. Horowitz, War and Peace in Contemporary Social and Philosophical Theory (London: Souvenir Press, 1973), pp. 68-87.

49. Collected Works, vol. 48, p. 407.

50. Harijan, vol. 9, p. 264.

51. See, for instance, his letter of November 1947 to Madame Privat reproduced in ibid., vol. 11, p. 453. More pointedly, "It has often been said that the doctrine of non-violence I owe to Tolstoy. It is not the whole truth, but

there again I derive the greatest strength from his writings."
Collected Works, vol. 48, p. 407.

52. Autobiography, p. 220.

53. Harijan, vol. 9, p. 264.

54. "The statement that I had derived my idea of civil
disobedience from the writings of Thoreau is wrong. The
resistance to authority in South Africa was well advanced
before I got the essay of Thoreau on civil disobedience. But
the movement was then known as passive resistance. As it
was incomplete I had coined the word satyagraha for the
Gujarati readers. When I saw the title of Thoreau's great
essay, I began to use his phrase to explain our struggle to
English readers. But I found that even civil disobedience
failed to convey the full meaning of the struggle. I, therefore, adopted the phrase civil resistance." Collected Works,
vol. 61, p. 401.

55. Ibid., vol. 45, p. 95.

56. H.S.L. Polak, H.N. Brailsford, and Lord Pethick-
Lawrence, Mahatma Gandhi (London: Odhams Press, 1949),
p. 58. Also see Speeches and Writings of Gandhi (Madras:
G. A. Natesen, 1933), p. 226.

57. Collected Works, vol. 48, p. 369.

58. Autobiography, p. 59.

59. Devanesan, Making of the Mahatma, p. 368.

60. Collected Works, vol. 47, p. 276.

61. Harijan, vol. 11, p. 87. Also, "Today [July 1947]
they were cutting one another's throats and they were preparing for further slaughter. If such a fight came, it would
be worse than during the Mutiny of 1857. In 1857, the
masses of India were not awake." Ibid., p. 263.

62. Collected Works, vol. 3, pp. 222-223.

63. Autobiography, p. 233.

64. Collected Works, vol. 9, p. 471; vol. 5, p. 56.

65. Harijan, vol. 10, p. 405.

66. Collected Works, vol. 29, p. 17.

67. Ibid., p. 17-18.

68. Harijan, vol. 11, p. 453.

69. Gandhi told Vincent Sheean (Lead Kindly Light), that orthodox scholars "have criticised my interpretation of Gita as being unduly influenced by the Sermon on the Mount" (p. 184).

70. Paul F. Power, Gandhi on World Affairs (Bombay: Perennial Press, 1961), p. 49.

71. All citations with regard to the Gita in this section are substantiated in Collected Works, vol. 15, pp. 312-313; vol. 21, p. 515; vol. 25, pp. 85-87; vol. 26, p. 335; vol. 28, p. 318; vol. 32, p. 417; vol. 35, p. 385; vol. 38, p. 319; vol. 41, p. 93; vol. 49, p. 69; vol. 63, pp. 260-262, 319-322.

72. Harijan, vol. 9, p. 40.

73. Sheean, Lead Kindly Light, p. 198.

74. Collected Works, vol. 22, p. 18.

75. Ibid., vol. 28, p. 320.

76. Leo Tolstoy, Kingdom of God and Peace Essays (London: Oxford University Press, 1960), pp. 214-220.

77. Collected Works vol. 25, p. 232.

78. P. Bhattacharya and A. Mukerji, Ruskin's Unto This Last (Calcutta: Alpha, 1969), p. 69.

79. John Ruskin, The Crown of Wild Olive, Four Lectures on Industry and War (London: George Allen, 1909), pp. 115-171.

80. Harijan, vol. 8, p. 250.

81. C. Seshachari, Gandhi and the American Scene (Bombay: Nachiketa, 1969), p. 19.

82. Collected Works, vol. 42, p. 437.

83. Ibid., vol. 48, p. 402.

84. The theme of Krishna's discourse to Arjuna is "Action alone is thy province . . . nor should thou desire

to avoid action." Mahadev Desai, The Gita According to Gandhi (Ahmedabad: Navajivan, 1956), pp. 128, 161.

85. Autobiography, p. 258.

86. Mahatma Gandhi, Non-violence in Peace and War, vol. 1 (Ahmedabad: Navajivan, 1948), pp. 24-25.

87. Sheean, Lead Kindly Light, pp. 58-61, 300.

88. Collected Works, vol. 18, pp. 443-444; vol. 63, p. 339.

89. George E. Catlin, In the Path of the Mahatma (London: MacDonald, 1948), p. 259.

90. Harijan, vol. 8, p. 266.

91. Ibid., vol. 10, p. 28.

92. Collected Works, vol. 63, pp. 321-322, 339.

93. Ibid., vol. 62, pp. 175-176.

94. Ibid. vol. 15, p. 289; vol. 35, p. 332.

95. Quoted in Kshitis Ray, ed., "Mahatma Gandhi and Christian Pacifism," in the Gandhi Memorial Peace Number of the Visva Bharati Quarterly (Santiniketan) 1959, pp. 137-141.

96. Autobiography, pp. 317-320.

2
THE ENTANGLEMENTS OF WAR

The years approaching the end of the nineteenth century saw southern Africa preparing for a showdown. Four colonies existed in the last decade of the century in that part of Africa: Natal, Cape, Transvaal, and the Orange Free State. These were ruled by the descendents of Europeans who had chanced upon the area centuries ago on their way to the fabled India. The Europeans settled and developed the area first as a convenient halfway house to the East, and later on as their home.

The Dutch, finding a welcome release from the cramped confines of their homeland, too, quite understandably, took to farming and enjoyed virtually undisputed rule over southern Africa for nearly 200 years. The British, arriving in the nineteenth century, wrested control of the Cape in 1806 and Natal in 1843 from the Dutch, who thereupon moved inland and brought Transvaal and Orange Free State under their direct control.

Thus, by the time Mohandas Karamchand Gandhi arrived in 1893, even though the British were dominant in the Cape and Natal, and the Dutch farmers (the Boers) were preponderant in Transvaal and the Orange Free State, the struggle between the two communities for hegemony over the whole of the southern Africa had sharpened to a point of no return. The root cause of the eventual clash between them was Transvaal, of course, with its huge deposits of gold. But the pretense for the opening of hostilities, which go by the description of the Boer War and lasted until 1902, was purported to be the reform of government, the grievances of the uitlanders (foreigners), and the ill-treatment of indentured farm labor from India at the hands of the Boers.

Gandhi, at the time the war broke out on October 11, 1899, we may remember, was deeply involved in the struggle of Indians against racial discrimination by the Europeans; the Indians were demanding protection by the British government of South Africa as British subjects. Gandhi saw in this an opportunity to rectify the general impression that "in time of danger the Indians would scuttle off like rabbits" and that besides being cowards they were despicable for always hankering after rights and privileges without ever bothering about responsibilities. In this time of crisis, the local Indian community of South Africa had to recognize and discharge its full responsibility as British subjects. Failing that, it would be at fault and even deserve the harsh treatment that was bound to come in the wake of the war. Gandhi's sole concern at that stage was to prove that the Indians were as capable as the Europeans or anybody else of making sacrifices and bearing responsibilities.

Thus Gandhi called a meeting of some English-speaking Indians of Durban at short notice within the week of the outbreak of the war to consider the desirability of unconditionally offering their services to the British authorities. The hundred or so who met heard him appeal that Indians forget their internal differences, shelve their opinion regarding the "justice of the war," and render service. He was aware that many of his countrymen held other views because of their unhappy and wretched lot, but he persuaded them to accept his view and succeeded in enlisting a few volunteers.

Before actually making the offer of the services of his compatriots to the British government, however, he consulted Mr. Jameson, a member of the Legislative Council well known to Gandhi. Jameson advised Gandhi against the move on the ground that the Indian volunteers were unable to handle arms; they should, in his opinion, offer financial help instead, which, incidentally, Gandhi had already done. In response to the complaint of William Palmer, treasurer of the Durban Women's League, that Indian merchants were not coming forward with donations to the league's war effort, Gandhi had promptly sent around a subscription list, which included his own contribution of three guineas.

But that was not the kind of help Gandhi felt to be adequate or commensurate with his main objective. So he talked to another friend, F. A. Laughton, who seemed to be in perfect agreement with Gandhi's entire approach: "That's the very thing," Laughton said, "do it; it will raise your people in the estimation of us all, and it will do them good."[1] Thus encouraged, Gandhi sent the formal offer of help to the colonial secretary. In his letter of October 19, Gandhi, while regretting that the lack of training in handling arms prevented him and his countrymen from fighting at the front line, urged that they could be of some assistance and help to the empire.[2] He assured the secretary that if unflinching devotion to duty and extreme eagerness to serve the sovereign could make them of any use on the battlefield, his countrymen would not fail. He suggested that perhaps they might render service in field hospitals and the commissariate and that they were quite prepared to work as orderlies, or simply doing the menial work in the hospitals, even as sweepers or scavengers, for the wounded.

The whole idea, according to Gandhi, was to prove that, in common with the other subjects of the Queen Empress, the Indians in her South African realm were also ready to do their duty on the battlefield; the offer was meant to be an earnest declaration of Indian loyalty. Acceptance of the offer would serve as a link binding still closer the different parts of the mighty empire, of which, declared Gandhi, "we are so proud."

The principal undersecretary conveyed the government's deep appreciation of Gandhi's offer, of course, but said that the government would avail itself of their services should the occasion arise! To Gandhi's great disappointment, this, in effect, meant that his offer had been politely but firmly declined. Was it the reluctance of a government chary of placing itself under the obligation of a former campaigner and civil resister?

Despite the government's rebuff, Gandhi's ardor to serve in the battle field was not diminished one whit. He talked to Dr. Canon Booth, who was in charge of the India Anglican Mission Hospital. Dr. Booth was pleased to find the Indians ready to serve the empire in the war and offered to train them as ambulance leaders. For weeks Gandhi along with many of his countrymen received training in first aid. At the same time, he kept on requesting the government to accept their services, but the government chose to respond with cold silence to these requests, until the war started going badly for the British in December 1899. The British commander, General Buller, finding that the available European ambulance corps just could not cope with the ever-mounting strain of casualties, asked the Natal government specifically to raise an Indian Ambulance Corps to meet the worsening situation, and the government decided to make use of the Indians' services. Seldom do people persist so doggedly in pressing their help, entailing danger and death, as Gandhi did.

The official government protector of Indian immigrants in Natal informed Gandhi that the government required 50 to 60 Indians to go to the front, who in turn enquired telegraphically on December 2, 1899 regarding the nature of work they would be called upon to do, and the date of commencement of that work. The colonial secretary still did not think the Indians were capable of doing ambulance work independently and asked only for stretcher bearers. Gandhi was sorely disappointed, for this meant that the Indian volunteers would virtually be laborers rather than those rendering medical assistance. On December 4, he informed the colonial secretary that his men were holding themselves in readiness to leave at a moment's notice. Regretting the government's tardiness in accepting them as capable of being medically useful, particularly when they had received the requisite training under Dr. Booth, Gandhi sent another list of volunteers. Dr. Booth endorsed Gandhi's claim and wrote to the colonial secretary about the work Gandhi's group could do. Gandhi, at the same time, urged the bishop of Natal to free Dr. Booth for ambulance work; accompanied by Booth he then called on the bishop and Colonel Johnston at Maritzburg. These meetings paved the way for acceptance of the fact that Booth and Gandhi would do admirably to act as leaders of the Indians engaged as bearers in the field. And so Gandhi was allowed to go to the battlefield, full of hope and confidence. His "dream was realised," though only partially since his countrymen "were not to be engaged in the fighting line."

Gandhi left for the front on December 14, 1899, at the time of the battle of Colenso; the mayor of Durban and other important Englishmen spoke words of encouragement to Gandhi and his contingent of volunteers. The group reached Chieveley the next afternoon at about 3:30, was given the red cross badges, and ordered to march at once to the field hospital about six miles away. Thirsty and hungry, they reached Colenso as the battle ended and were promptly ordered to carry the wounded to the field hospital where the European Ambulance Corps was already busy. They kept working until the medical officer asked them, at about eleven o'clock, to retire for the night; it was past midnight before they could get their first meal.

Work was resumed early the following morning and continued till about 11 A.M., and hardly had the work of removing the wounded finished when orders were received to break up the camp and march to Chieveley immediately to entrain there for Estcourt. It was a retreat, of course, but Gandhi was quite impressed by the efficiency with which 15,000 men, with heavy artillery, broke the camp.

It was an extremely hot day for marching, and that part of Natal had neither trees nor water, but they marched under these trying conditions for about three hours. On reaching the railway station at Chieveley, they were informed about the uncertainty regarding the availability of the carriages. At about 8 P.M. the European Ambulance Corps was put on the train to Estcourt, and the Indians were asked to spend the night on the veldt:

> Tired, hungry and thirsty--there was no
> water available at the station except for the

patients and station staff--the men had to
find means of satisfying both hunger and
thirst, and of obtaining some rest. They
brought dirty water from a pool about half
a mile from the station, cooked rice and by
midnight, after partaking of what was, under
the circumstances, regarded as an excellent
repast, wanted to sleep. Practically the whole
of General Buller's Cavalry passed by during
the night, and the men had very little rest.[3]

On the next day, December 17, the Indian volunteer corps
was closely packed in open railway trucks, and after about
five hours in the blazing heat the train proceeded to Estcourt. They reached Estcourt and the vagaries of inclement
weather. After waiting there, without shelter against hot sun
and strong wind, for two days, they were disbanded. Colonel Gallwey personally thanked them for the good work done,
told them to expect a call soon, and said that he looked
forward to a similar response when he needed them again.
General Wolf Murray officially recognized their services rendered during the retreat from Colenso.

The Indian Ambulance Corps had, in total, rendered
service for seven days. Gandhi and the leaders of the corps
had not accepted any money for the work, of course. But
they were pained at the discriminatory attitude of the British
government: the rates for Indian stretcher bearers were a
little over half those of their European counterparts, and
Indians had to sleep out in the open since no tents were
provided for them.

Meanwhile, General Buller was taking his men across the
Tugela to force his way to Ladysmith through Spionkop.
Within ten days, on December 29, 1899, Gandhi received a
letter from Colonel Gallwey, the principal medical officer of
the Natal militia, enquiring as to the number of Indians
available for work as stretcher bearers for the battle of
Spionkop. Gandhi replied telegraphically that 500 Indians
were ready to work as before until the end of the war. When
he reached Estcourt on January 7, 1900, Gandhi had over
1,000 volunteers with him, and the Indian Ambulance Corps
was re-formed--this time under a somewhat better and more
auspicious conditions, probably also because Gandhi and his
corps of volunteers had impressed the authorities with their
work some days ago. Now they were to be provided with
tents and were not to work within the firing line. The re-
formed corps, however, had to while away a fortnight drill-
ing under Dr. Booth, till they were ordered to entrain for
Frere, and from there they marched 25 miles to Spearman's
camp near Spionkop; the march was not without its trials
and tribulations.

The field hospital, where the Indian corps had to work, fell within the range of fire. Although the Indians were not supposed to work within the firing line, emergency circumstances prompted Major Bapty, secretary to Colonel Gallwey, to request that Gandhi and his men go to the field hospital since a large number of wounded were to be removed from there. Gandhi agreed and the corps gave an excellent account of itself. From January 21, the corps was incessantly at work for three weeks: with its precious load of wounded, it covered a distance of 25 miles per day three or four times, which was something of a record as Gandhi recalled later.

Once more at Vaalkrantz, Gandhi worked within the range of fire with shells falling just yards about him as he helped remove the wounded. The Indian Ambulance Corps was disbanded again on January 28, 1900. Though the Boer War dragged on until 1902, Gandhi's services were not requested after this disbanding.

Gandhi's action in organizing the Indian Ambulance Corps during the Boer War, it must be realized, was a perfectly moral, expeditious, and a typically Gandhian solution to the difficult dilemma of whom he should support: the Boers, whose case on his own admission enjoyed his sympathy, or the British, to whom he was committed by virtue of his declared loyalty to the Queen and the British Empire. Serving the suffering and the wounded appeared to him to be the answer. Moreover, his persistent offer of help again was in accordance with his principle of satyagraha. He was acutely conscious of the hatred of the colonials against the Indians and was actively engaged in organizing a struggle against this intense hatred. But when the war came, he suspended all confrontation, considering it to be "prudent to stay further action till the clouds melted away."[4] As a true satyagrahi he decided not to turn the opponent's hour of need into his opportunity; he would conquer that "hatred by love."

Gandhi was drawn to the battlefield for the second time in 1906, when in protest against a new tax levy by the Natal administration in February, the Zulu tribesmen came out in open rebellion against the British government.

He read about the Zulu Rebellion in the papers, but for some weeks it hardly made any impression on him. As he realized that the trouble was persisting, Gandhi, on March 17, 1906, appealed to the government to allow the Indians help in bringing about an end to the conflict.[5] The British government ignored his appeal, of course. He reiterated his offer again toward the end of the month and wondered about the lack of response from the government. Getting no reaction from the government, Gandhi, at the end of April, convened a meeting of the Natal Indian Congress to consider the advisability of raising an Indian Ambulance Corps. At the meeting he exhorted the Indians to offer to the government their assistance irrespective of the opinion they held in regard to the cause of the rebellion. Their personal disabilities or limitations ought also not make them either diffident or prejudiced; they should, indeed, do their duty of averting the danger threatening the colony. The meeting adopted a resolution to the effect that Indians, forgetting for the moment their personal grievances, would help in ending the rebellion. It was well publicized in the press, and many ridiculed the idea of Indians going to the front line. But Gandhi could not be more serious: he urged the government to train Indians in the art of warfare. After all, where was the wisdom, or statesmanship, in blindly refusing to make use of 100,000 Indians resident in the colony for the purposes of war?

Here was a subtle way of getting the Indians a status of equality with other residents of the colony. Gandhi believed by their service in the war the Indians would be able to obtain redress for the wrongs done to them. Indeed, now was the time for them to prove their mettle. And he was quite candid about his belief, let alone feel apologetic about it. He assured his countrymen that, if they wanted to live happily and with dignity in Natal, they should always be ready to participate in war without fear or anxiety; and they were eminently qualified to do so. They had profound faith in God, a greater sense of duty, which makes it easier for them to volunteer; death did not scare them for hundreds of thousands die of famine or pestilence in their native country; a long tormented drawn-out existence ending in death was no better than the death that might overtake one on the battlefield. Moreover, during the Crimean War and during the attack on Ladysmith in the Boer War, Gandhi reminded the Indians, fewer men died from bayonets and bullets than through disease, accident, and sheer carelessness or perverse living. It was, therefore, glorious to die on the battlefield.

The colonial administration, at last, accepted Gandhi's offer on May 30, 1906, when he was asked to organize an Indian Ambulance Corps of about 20 stretcher bearers.[6] He at once communicated to the government the readiness of his men and placed the volunteers at the disposal of the principal medical officer in Natal. He informed Dadabhai Naoroji and Gopal Krishna Gokhale that in view of the duties connected with the ambulance work at the front, the deputation to England had been held in abeyance.

In contrast to his overtures and association in the Boer War, Gandhi was a little more demonstrative in the Zulu Rebellion. Whereas at the time of the Boer War he had called upon the Indians to volunteer somewhat quietly, and perhaps without much hope of being accepted, he now called for re-

cruits for combat; he now seemed to regard serving as a
stretcher bearer as the very last resort. But since the government remained unconvinced that Indians would make good
recruits, he had to content himself with raising a small ambulance corps of 20 men, a third of them being Tamils, another third Muslims, and the rest Gujaratis and north Indians. They solemnly took the oath of allegiance to King Ed-
ward VII, his heirs and successors, and agreed to the terms
of service, which included rations, uniforms, equipment, and
one shilling and six pence a day. Gandhi was given the rank
of sergeant-major. There were three sergeants, one corporal,
and 15 privates under him. The group in their field uniforms
makes an interesting photograph!

Marching orders were received on June 21, 1906, and the
Indian volunteers were on the train to Stranger next day to
join Colonel Arnott's forces. On arrival they did not find any
sign of Zulus, but Arnott's men afflicted with malaria or
those who had had accidents engaged the attentions of the
volunteers. They had to rough it in that they had to sleep
on the ground in the open just in their overcoats and a
blanket. It was cold but Gandhi fed his men before they
went to sleep.

A full-kit march to Mapumulo followed; it was an un-
eventful treck except for a raid on an orchard and later the
happy discovery that they could carry the stretchers and
medicine chests in the wagons. An uphill journey of two days
brought them to Mapumulo, and their first Zulus. These
Zulus were poor peasants and farmers who had been rounded
up by the troops and thrown into a stockade. They bore the
marks of beatings, which had been administered to the poor
wretches just as warnings to them and their brethren not to
rebel in the future. They lay in filth and misery, often in
their own blood in the stockade. No English doctor would
tend their wounds, and the task was given to the Indian Ambulance Corps. A makeshift hospital was set up, and soon
Gandhi and his colleagues were tending to the suffering
Zulus while the British troops, which had administered the
beatings, jeered from behind the railings of the stockade.

An advance party of stretcher bearers under Gandhi was
then sent up the line to Otimati, where an engagement was
thought imminent. There was, however, no engagement. A
British trooper had crushed his toe under the wheel of a
wagon, and another had been accidentally shot in the thigh.
Gandhi was ordered to carry them on stretchers to the base
hospital. The terrain was rugged, the men on the stretchers
heavy, and the Zulus observed from the hills. Suddenly it
occurred to somebody in the British command that the Zulus
may have gotten the impression that they had succeeded in
inflicting injury on the British forces. So off came the men

from stretchers on to the ambulance wagon, and the bearers heaved a sigh of relief.

A more important assignment was given to Gandhi a few days later, when his group was ordered to accompany a flying column of mixed cavalry and infantry. There was to be no ambulance wagon for carrying the heavy equipment. The volunteers had difficulty in keeping pace with the column and during the night lost contact with it. When dawn came they found themselves wandering in the hills, with not the slightest idea as to their bearings. The sudden appearance of a Zulu with a weapon known as an assegai frightened them, and they apprehended being slaughtered. But it was perhaps a lone Zulu with none of his comrades in the hills. Eventually the volunteers caught up with the column and soon got busy in tending to the Zulu "friendlies" accidentally shot by the troops.

It was a strange little "war," with enemy melting into the landscape, and the British forces, consisting almost entirely of volunteers, behaving like utter amateurs, scarcely knowing what they were doing and what needed doing. Duncan Mckenzie, who commanded these troops, was a farmer in private life. Colonel Wylie was a well-known Durban lawyer, and Colonel Sparks was a butcher by profession. Gandhi had known both the colonels in 1897 as members of the committee protesting against invasion of Indian laborers on the S.S. court land. They seemed changed men now, more gentle, and they went out of their way to say nice things about the Indian stretcher bearers.

Gandhi's most unforgettable memories of the "war" were those of endless marches, the thirst, the cold nights, the flogging of Zulus at Mckenzie's orders, of how they sometimes marched 40 miles in a day, and how they seemed to be continually crossing and recrossing the winding Umvoti River. Sometimes the friendly Zulus were ordered to assist Gandhi's team, and he was puzzled to find that these Zulus hardly took any interest in their wounded brethren: "The natives in our hands proved to be most unreliable and obstinate. . . . Without constant attention they would as soon have dropped the wounded man as not, and they seemed to bestow no care on their suffering countryman."[7]

Often Gandhi found time hanging heavy, for there was no visible front on which the services of his team could be focused. So he sent short notes about the conditions of work and his team, detailing, for instance, the exact amount of bread, sugar, tea, and other rations given to his men; the Indian Opinion published these as coming from "Our Special Correspondent At The Front"! More important perhaps, the long marches through the countryside gave Gandhi a good deal of time for reflection and introspection, which led to a personal achievement that brought about the most far-reaching change in his life: this was his decision to embrace brahmacharya (sexual abstinence). The realization deepened in him that he could possibly not continue to lead two lives, so to speak, namely, a life of service and of marital enjoyment. He then opted for a total dedication to service.

After the Indian Ambulance Corps had worked for about six weeks, it was disbanded on July 19, 1906. The Natal Indian Congress gave a reception to the returning team, at which Gandhi suggested that Indians should be allowed to join the permanent volunteer corps. Another meeting organized by the Congress resolved to decorate the members of the disbanded corps with medals. Gandhi pursued with the principal medical officer of the Natal militia his idea of instituting a permanent ambulance corps. At the same time, he demanded special training and arms for self-protection for Indians since they were untrained and untried.[8]

The third, and last, time that Gandhi assisted in war was from 1914 to 1918, and it was a final expression of his loyalty to the British Crown, strongly felt and stubbornly flaunted. He sailed for home in 1914 and never revisited South Africa. En route he halted in London and discovered that England was at war. The "tremendous crisis which had overwhelmed the Empire" set him thinking as to what was his duty under the circumstances.[9] As his Indian and English friends honored him by a formal reception, Gandhi gave expression to his earnest desire to do his duty as a citizen of the empire, even though his views on the empire had not changed. He fought the British Government with every weapon available to him whenever necessary, but when the empire was in peril as a loyal subject of the Crown, he would offer his services to the cause of its survival. So he decided to make the offer to the War Office.

In a circular letter to his acquaintances in London, he suggested that they place themselves unconditionally at the disposal of the authorities.[10] In his opinion, under the given circumstances, whatever be assigned to the Indians should not be considered beneath their dignity, or inconsistent with their self-respect; hence his deliberate use of the word unconditionally. Response to the circular was favorable, and so he wrote to the undersecretary of state for India that his fellow Indians desired "to share the responsibilities of membership in this great Empire, if we would share its privileges."[11] Since Gandhi had offered to serve in any capacity, the India Office assumed that he wanted to raise a field force that would eventually be incorporated in the army. There were long confabulations in the government until Lord Crewe--the secretary of state for India--concluded that it would not be advisable to enlist in military service the Indian students residing in Britain since it would make them neglect the purpose for which they had come to England;

instead, the Indians could serve in a field ambulance corps.[12]
Gandhi saw the point.

Soon, about 60 Indians were taking Red Cross training
at the Regent Street Polytechnic, with Dr. James Cantlie as
their chief adviser. Thus Gandhi, who barely five years
earlier had spoken of his horror and detestation of doctors
in Hind Swaraj, found himself a willing pupil of that gruff,
kindly doctor.[13] The six-week training course in first aid
was to be followed by a period of military training and then
some more first aid at Netley, a village in Hampshire. Gandhi quite happily went about learning and qualifying in
things like: "What is the cure for opium poisoning?" "What
is the treatment for a broken collar bone?" "What should be
done to stop bleeding from a wound in the palm?" And so on.

In the meantime, Gandhi also issued a general circular,
on September 22, 1914, inviting more volunteers. At a public
meeting held at the Polytechnic, on October 1, and attended,
among others, by the Agha Khan, Gandhi spoke of Dr.
Cantlie's splendid spirit in training the Indians and presented the doctor with a set of Tagore's works in appreciation of
his services to the corps.

Some problems arose, however, at the Netley training
camp. The Indian volunteers at the camp had been placed
under military discipline, but they liked to take their orders
only from Gandhi, the chairman of the volunteer corps. The
camp commander strongly resented this, however. So when
he appointed corporals from among the ranks of the privates,
the Indians objected, demanding that either Gandhi performed
that function or they would themselves elect the corporals.
While supporting the position of the Indians on the issue,
Gandhi pointed out to the commandant, Colonel Baker, that
this was the practice they were used to. In the Boer War
and the Zulu Rebellion, the commanding officer had never
interfered in the election of the section officers or in the
internal administration of the corps. A good deal of correspondence flew back and forth between Gandhi, Colonel
Baker, and the undersecretary of state. The Indians threatened to go on strike and Gandhi concurred. A meeting held
October 13, formally adopted a resolution to this effect. Gandhi communicated to the colonel:

> If you could possibly see your way, in the
> interest of the Corps, to alter your decision,
> the drilling will go on. If you could not do
> so, those of us who voted for the resolution
> and others who may fall in with it will be in-
> formed of your unfavourable decision and will
> therefore respectfully withdraw.[14]

Here was satyagraha being applied in a quasi-military situation, and Gandhi was not very pleased in having to resort to it.

Suddenly, an unexpectedly large number of wounded soldiers arrived at Netley, and there was no longer any question of disobeying military orders. The undersecreatry of state, in a cautiously worded letter of explanation, urged Gandhi to place the Indians under the command of the authorities at Netley so that the wounded would receive due aid. Gandhi acceded to the request, and the volunteers resumed work. While deriving some consolation from the fact that he had practiced a miniature satyagraha in wartime, Gandhi had found the situation "unfortunate," "exhausting and exasperating;" the arrival of the wounded had saved the situation. And once again, he started making appeals to Indians to volunteer since more and more volunteers were needed.[15]

Gandhi was not well at that time. He suffered a prolonged bout with pleurisy primarily because he insisted on being his own doctor. He pushed away all thoughts of leaving for India, for he felt his duty lay in England till the end of the war. And he was a prey to the nagging doubts regarding his actions: Was the forming of the volunteer corps the right thing to do? Had he been right in demanding unconditional cooperation of the Indians? Everyone knew, of course, that he was in no state of health to bear arms, but should other Indians engage in fighting? And so on. These months in wartime England were among the most terrible Gandhi had ever had in his life. Never again would he be so conscience-stricken, so vulnerable. Even while lying sick in bed, he made plans to go to Netley to nurse the wounded.

The undersecretary of state, Charles Roberts, and his wife were concerned about Gandhi. Seeing the state he was in, they advised Gandhi to go to India immediately. Gandhi, of course, argued against such a course: surely he could be of use in some other position if not tending the sick at Netley; there must be some way in which he could be of service to England. But Roberts was insistent that Gandhi had done enough service to the Crown and that nothing more could be demanded of him, in fairness, except that he recover his health, which he could in India. Gandhi gave in reluctantly. His passion to serve the British Empire came through, however, even in his farewell speech where he made clear that he had not resigned from the volunteer corps and that he intended to come back when he regained his health should the hostilities still exist.[16]

Gandhi thus returned to India and continued in whatever way he could to support the war effort; more relevant is the fact that this time it was the government that repeatedly approached him for help.

The war entered a critical phase in early 1918, and the Allies desperately needed help. In order to harness manpower, Britain turned to India: the princely India had already committed large contingents of men to the war, but the British India could certainly do even better. Lord Chelmsford, the viceroy, convened a war conference at Delhi toward the end of April, and Gandhi was invited to it. He did attend on the first day, April 25, but for a number of reasons refused either to serve on any of the committees or to speak on the main resolution, which read:

> That this conference authorises and requests
> His Excellency the Viceroy to convey to His
> Majesty the King Emperor an expression of
> India's dutiful and loyal response to his gra-
> cious message, and assurance of her determi-
> nation to continue to do her duty to her
> utmost capacity in the great crisis through
> which the Empire is passing.[17]

One of Gandhi's reasons for refusing to participate more actively was that the government had excluded most of the important Indian leaders, such as Tilak, Annie Besant, and the Ali Brothers, and their absence, in Gandhi's opinion, rendered the conference quite meaningless. Another reason was Gandhi's discovery of the secret treaties that Britain alleged- ly entered with Russia in regard to Constantinople.[18] Gan- dhi was not sure that in the light of this discovery he would still hold the cause of the Allies to be just and worthy of sympathy.

However, on April 27 he met Lord Chelmsford who na- turally repudiated the report about the alleged treaties as having emanated from interested sources. After this assur- ance, Gandhi considered it to be a matter of duty to join the conference--though "in fear and trembling." He told the con- ference members that he considered himself honored to be among the supporters of the resolution about recruiting: "I realize fully its meaning and I tender my support to it with all my heart."[19] To the viceroy he wrote:

> I recognise that, in the hour of its danger,
> we must give--as we have decided to give--
> ungrudging and unequivocal support to the
> Empire, of which we aspire, in the near
> future, to be partners in the same as the
> Dominions overseas. But it is the simple
> truth that our response is due to the expec-
> tation that our goal will be reached all the
> more speedily on that account.[20]

He further promised to make India "offer all her able-bodied sons as a sacrifice to the Empire at its critical moment;" but while he was willing to sacrifice "every available man," he could not offer any financial assistance. He urged that after the war Indians be made equal partners in the empire, cautioning, at the same time, that "disappointment of hope means disillusion." He was ready to do anything the viceroy considered real war work; the nearest would be recruiting for war, as he himself would not wield arms. "I shall not commence work," however, "before I have your reply." Clearly, Gandhi was convinced that, if India participated in the war on the side of the British, it would obtain swaraj (self-government) within less time and with less effort. So he went ahead with preparing himself for the task he had undertaken for the British.

He started saying that India should provide men and impose no conditions; any calamity that overtakes the empire is one that overtakes India as well; Indians should not let this opportunity slip for it would never come again; they should get military training: "We cannot have swaraj until we have made ourselves fit for it. One aspect of this fitness is surely that we should share its burdens. . . . The end of the war will see us better qualified [for swaraj]."21

A couple of months later (June 10, 1918), Gandhi was invited to a war conference in Bombay. After one look at the prospective participants, he promptly declined the invitation since men like Tilak had not been included among the invitees. Gandhi feared that his "usefulness will be materially curtailed if [he] could not have the benefit of [Tilak's] cooperation and that of other Home-Rulers of his calibre. It is hopeless to expect a truly national response and a national army, unless the Government are prepared to trust them to do their duty." He would serve on the manpower committee only if other Indian leaders were included in the proposed conference. Lord Willingdon, the governor of Bombay, was quite startled at Gandhi's response and had him reminded that the latter had offered unconditional cooperation in the war effort. The governor then requested a meeting with Gandhi.

The issue was resolved at the meeting, and Gandhi started conferring with his close associates regarding recruitment. At the same time he suggested to the British prime minister that full mobilization of manpower would be feasible and vastly fruitful if the government adopted an attitude of trust toward the people of India. As indicators thereof, the Arms Act should be amended forthwith to enable people to carry arms; Indians should be given the opportunity to become officers and hence admitted to the ranks of the King's Commission; and a military college for training Indian officers be started as soon as possible.

Gandhi wanted to take in the whole of the country at once for his campaign of recruitment for the British, but started with the district of Kheda in Gujarat, formally on June 21, 1918. He was familiar with the area, for only months ago he had conducted satyagraha for its peasants against government oppression and persecution unleashed in the wake of crop failure and the consequent inability of peasants to pay land revenue.

The main thrust of his argument in the entire campaign can easily be summed up in one sentence: cooperation with the British in the war will save the empire and will be rewarding in the attainment of self-government while training the Indians in the art of self-defense.

Thus, Gandhi maintained that in the context of the ongoing war the English were superior to the Germans; he was himself prepared to go to war provided the people joined him. He was acutely conscious of the government's failings, of course, but the people should hold their grievances in abeyance for the time being and put in a spirited effort to meet the threat to the empire. After all, the main objective of the people of India was "partnership in the Empire." This could be achieved if Indians bore utmost suffering, even laid down their lives in defense of the empire; for if the empire went down, their cherished aspirations would perish too.

Gandhi was confident that Indian sacrifices for the empire would not be in vain; the British, in recognition of the heroism and services of the Indians, would reward them duly by granting them home rule. Moreover, said Gandhi, it was of utmost importance that the Indians shed their fear of the military. Not only was military training the stepping-stone to home rule, it behooved the people of India to learn the use of arms and acquire the ability to defend themselves --for how long would Indians remain dependent upon the British for their own defense? So if they wanted to learn the use of arms with the greatest possible dispatch, it was their duty to enlist in the army. By availing themselves of the "golden opportunity" provided by the war, the Indians would, in fact, be acquiring the capacity of self-protection and self-defense.

Lastly, Gandhi exhorted that by enlisting in the war the Indians would convincingly put an end to the impression that they were cowards. They could banish the fear of death from their hearts, for death is in any case inevitable. Pestilence and epidemics claimed hundreds of thousands of lives in the country, but death as soldiers on the battlefield would make them immortal. It was rumored that Indian soldiers bore the brunt of the enemy attack since they were always pushed on to the front line to do the fighting, bear the bullets of the enemy and get killed, while the British stayed behind.

This, Gandhi assured, was a canard, for he believed the British to be a valiant and fighting race that would not take shelter in the rear. If the rumor were substantiated, however, Gandhi himself would strenuously object to Indians being pushed to such vulnerable positions, even at the expense of being shot himself.

In sum, then, Gandhi argued that Indians should establish their bravery and merit in order to be worthy partners in the British Empire. In order to be really free, they should learn to use arms; and the administrative follies and lapses of the British notwithstanding, they should put their full faith in the British, as he himself had done.

It would be wrong to consider all this as a deviation on Gandhi's part from his natural, basic position on war. However strongly he felt about the crisis of the British Empire out of loyalty for the institution--and his feelings for the empire were strong and transparent for all to see--there is no question that the crisis did not weaken in any way his fundamental faith in nonviolence.

The rhetoric of Gandhi's campaign and canvassing has to be seen in the perspective of his sincere belief that swaraj would belong to India if it could convincingly impress upon Britain in its hour of peril India's worth and sacrifices. Freedom of the country was the supreme, overriding goal; thus his personal views on war could be temporarily kept in abeyance, especially when his duty as a citizen also demanded so. We have noted already how before reaching India he was assailed by doubts about his volunteer corps and particularly in insisting on the unconditional cooperation of his compatriots in England. It is difficult to accept that on reaching home his conscience was suddenly assuaged enough to make him a warmonger of sorts. In our view, the tone and tenor of his recruitment campaign were directly related to the prevailing atmosphere in the country of frustration, resentment, and mistrust over British intentions and policy. In such a noncooperative, even hostile environment, he fashioned his appeal in the style of an advocate arguing his brief, keeping his personal views aside for the moment.

Moreover, his appeal for cooperation with the British government was, evidently, addressed as much to the leaders of the Home Rule League. Tilak, Annie Besant, Mohammed Ali Jinnah, and others neither shared his faith in the British nor his enthusiasm for helping them in their crisis; they differed with him sharply. Tilak, for instance, insisted that the offer of Indian help should be made subject to certain clear conditions, thereby obviating the chance of any subsequent misunderstanding arising once the war had concluded. He offered to recruit 5,000 from Maharashtra and even sent a check for Rs. 50,000 to Gandhi as an earnest sign of his cooperation. But at the same time Tilak insisted that

Gandhi secure a promise from the government that, for instance, Indians would get commissioned rank in the army. Gandhi declined to make Indian help in the nature of a bargain and returned the check. So he persuaded the other leaders, but without being able to touch these leaders with his faith in the British or bringing them round to his viewpoint. Thus in a sullen atmosphere marked by noncooperation, Gandhi had to go it alone.

He started his recruitment campaign with the hope of raising nothing less than 1,000 recruits from Gujarat, two persons in 200, or 20 from each village. But almost a month's drive did not produce even a single recruit. Wherever he went he was greeted with cold incomprehension and ill-concealed hostility. These were the very people who, not very long ago during the Kheda satyagraha, had supplied his team with volunteers, transportation, and whatever goods they could. Now they would not lift a finger or lend him their bullock carts even when he offered to pay. He was reduced to trudging about the district with a handful of loyal but disspirited disciples, carrying their food in satchels. Villagers did turn up at his meetings, but only to heckle. How could a preacher, and a believer of ahimsa, exhort them to take to arms, and what had the government ever done for them, they asked?

Thus, despite dogged persistence and walking sometimes 20 miles a day, Gandhi was able to raise barely 150 recruits for the king and the empire by about the middle of August 1918. Would the people accept his advice only when it suited or benefited them, he wondered? On August 11 he was taken seriously ill and had to be shifted from Nadiad to Ahmedabad. This time, curiously, he did not quite regret that his illness had stood in the way of his services to the empire; but then the Allies were scoring victories on the battlefield every day now, and pressure had eased for Britain. On recovery, he held himself in readiness for going to wherever he might be needed, whether it be France or Mesopotamia. But with the German surrender he got word from the viceroy that his services for recruitment of soldiers were not necessary. He was now free of what he had seen as his obligation. Hardly had the armistice been signed when Gandhi declared, on November 14, 1918, that he was starting his boycott of foreign goods. His message was delivered at the opening of the Swadeshi Store, an outlet for selling goods manufactured in India, which were essentially rurally produced items. Apparently, Gandhi had his own ethics of cooperation!

THE NATURE OF GANDHI'S
INVOLVEMENT IN WAR

In all the wars he had been exposed to, Gandhi never wielded a weapon; his participation was thus of noncombatant nature. His assignments were to look after the wounded soldiers, a purely humanitarian work. A photograph of his, taken at the time of training under Dr. Booth as a member of the ambulance corps, shows him sitting next to the doctor. His biographer, Robert Payne, thinks that Gandhi in the photograph does not look like a successful lawyer but more like a man who has spent his life in the Red Cross and is heavily burdened with responsibilities: the eyes are watchful, and he is clearly anxious that men give a good account of themselves.[23]

Despite the noncombatant nature of his service, Gandhi's enthusiasm was tantamount to that of a warrior. In fact, he seemed to have been enjoying a certain kind of satisfaction in his assignments. A correspondent of the Pretoria News, who found Gandhi sitting by the roadside after a night of stretcher bearing and contentedly munching a regulation army biscuit, reported: "Every man in Buller's force was dull and depressed and damnation was heartily invoked on everything. But Gandhi was stoical in his bearing, cheerful and confident in his conversation and had a kindly eye."[24]

Twice during the Boer War, Gandhi was on the front serving the wounded. Once, at Spionkop especially, his duties were the more arduous and fraught with danger insofar as he had to work within almost 100 yards of the Boer guns with shells falling about him. Then he had to carry the dying Major General Woodgate to the base hospital at Frere; the agony of the general on the stretcher during that march was excruciating. Gandhi hurried through the heat and dust, fearful lest his charge should die before reaching the hospital. This was not the only experience of its kind Gandhi had during that time. But his physical courage was never in doubt; indeed, it was quite outstanding. Thus it is not surprising that his role and service left a deep imprint on everyone. General Buller praised him in dispatches, and the British government took due note of his contribution. Sir John Robinson, the first prime minister of Natal, spoke of Gandhi thus:

> I cannot too warmly thank your able country-
> man, Mr. Gandhi, upon his timely, unselfish
> and most useful action in voluntarily organizing
> a corps of bearers for ambulance work at the
> front at a moment when their labours were
> sorely needed in discharging arduous duties
> which experience showed to be by no means

devoid of peril. All engaged in that service
deserve the grateful recognition of the commun-
ity.25

Gandhi, however, never claimed any credit for himself.
He was happy that Indians had proved the British impressions
about them to be false and that he had served the empire in
its hour of need. Indeed, he often protested on being singled
out for compliments for the services rendered by his team.
His letter of March 30, 1901, to the colonial secretary is a
testimony in this regard, for it protests his being singled out
for mention in General Buller's dispatches. He always maintained that, if he were at all entitled to any credit for having
done his duty, his coworkers were equally entitled to it, for
without their cooperation he would not have been able to be
of any service.

Gandhi celebrated the achievement and success of the
British commanders in the Boer War. Not only did he send a
telegram of felicitations to the colonial secretary on General
Buller's relief of Ladysmith and ultimate victory, but also
convened a meeting for congratulating the other generals.
The invitation for the meeting bore the heading "Long Live
Kaiser-i-Hind" along with Empress Victoria's picture, as well
as the pictures of three prominent British commanders who
had served in the war. The meeting was held at Durban on
March 14, 1900. In his speech Gandhi said that the Indians
were particularly proud of the victory of Lord Roberts and
Sir George White--for many years the commander in chief in
India--who conducted the siege of Ladysmith, with such gal-
lantry; and that the Indians would have failed in their duty
to themselves had they not "given expression to their feelings at the success which had attended the feats of the two
Generals."26 Resolutions to this effect were adopted at the
meeting and given publicity in the press; the generals personally wrote to Gandhi in acknowledgment of these felicitations.

Gandhi was very proud, indeed, of the contribution that
the Indian community had made to the Boer War and was
eager to pen it down promptly. He restrained himself, but
only briefly, however, when an English friend advised that
"their part was merely to do without speaking." But after a
while, even as the war was on, Gandhi could not contain
himself any further and wrote, in March 1900, a detailed account of the activities of the Indian Ambulance Corps for the
Times of India. On many other occasions, he very proudly
narrated his experiences of the war and his contribution
therein; even after a couple of years he could accurately
recall and recount his experiences of that time. His pride in
this regard flowed from his conviction that ambulance work
was "as useful and quite as honourable as the shouldering

of a rifle" and just as necessary as the digging of trenches. Indeed, he considered his participation in the war to have been that of a "civilian soldier."

Apart from sending presents to the stretcher bearers, Gandhi wrote personal letters of appreciation and gratitude to the leaders of the ambulance corps.[27] "I have a feeling," he wrote, "that you enlisted partly out of regard for me and to that extent I am beholden to you." He could not possibly compensate them for their help and regard in terms of money, but he was certainly able to offer them, and their friends, his legal services in Durban free of cost up to £5 for a year starting April 20, 1900.

His pride in their work--and in the fact that the Indians during this war had proved their physical capabilities as well as their reliability to stand by the empire even at the peril of death--took him a step further. Gandhi urged the British government not to allow this enormous reserve force to go to waste out of sheer prejudice, but, instead, to give the "Indians the opportunity of a thorough training for actual warfare."[28] This, he pointed out, could easily be done under Law 25 of 1875, specifically enacted to increase "the maximum strength of the volunteer force in the Colony by adding thereto a force of Indian Immigrants Voluntary Infantry;" the governor was authorized by the act to "accept, with the consent of the employer, the services of any Indian immigrants who may be willing to be formed into a Volunteer Corps." Thus, "a very fine volunteer corps could be formed from Colonial-born Indians" that could serve Natal efficiently not only in peace but also in actual war.

For his services, Gandhi, along with 37 of his comrades, was awarded the War Medal. Volume 3 of the Times History of War in South Africa records:

> Mr. Gandhi, a lawyer whose persistent
> advocacy of Indian interests had brought him
> no little popularity, saw that now was an
> opportunity for justifying their demands for
> privilege by giving an example of patriotic
> duty. . . . Volunteer Ambulance Corps
> proved itself of great use. . . . [p. 100][29]

In the Zulu Rebellion, Gandhi was quite skeptical about the "rebellion" part of it. But as an ardent believer that the British Empire existed for the welfare of the world, he offered his services in the British cause. As Tendulkar remarks, "on reaching the scene of the 'rebellion', Gandhi realised that it was in fact a no-tax campaign. His sympathies now were with the Zulus. . . . The 'rebellion' was an eye opener to Gandhi. He saw the naked atrocities of the Whites against the poor sons of the soil. . . ."[30]

We are thus inclined to view Gandhi's services at the time of the Zulu uprising as something of a precursor, or a variety, of satyagraha but with a difference. The plight and the position of the Zulus made him sympathetic to them, but his obligation as a citizen of the empire pulled him to discharge his duty. The call of duty came first, and the discovery of the real issue at stake in the so-called rebellion followed his arrival at the actual scene. By continuing with his role in the bloody imbroglio, Gandhi met the compulsions of his conscience. In rendering the hapless Zulus selfless and sincere service, he signified protest while at the same time discharging his duty toward the empire.

In any case, the nature of his participation in the Zulu campaign was similar to that in the Boer War, except that now besides carrying the wounded other duties were also pressed upon him. Along with the other volunteers, he was called upon to disinfect the camps and dress the wounded. He was the compounder and dispenser both to the whites and the natives. Nursing of the individual Africans was Gandhi's singularly invaluable service, for the whites showed not the slightest interest or inclination to minister to the wounded Zulus. Quite possibly, without the Indian stretcher bearers, the hurt Zulus would have been left to die. In addition to these services, Gandhi also appealed to Indians to contribute to the Soldiers' Fund collected by the Durban Women's Association. The reward to Gandhi for the services rendered was the honorary rank of sergeant major.

Gandhi's active participation in the effort for World War I is characterized by a somewhat noticeable revision of his earlier stance. His letter to Maganlal Gandhi is an eloquent testimony to this.

> Recently, I used to say in South Africa that as Satyagrahis we cannot help in this way [i.e., by nursing the wounded] either, for such help also amounted to supporting a war. One who would not help a slaughter-house should not help in cleaning the butcher's house either.
> But I found that, living in England, I was in a way participating in the War. London owes the food it gets in wartime to the protection of the Navy. Thus to take this food was also a wrong thing. There was only one right course left, which was to go away to live in some mountain or cave in England itself and subsist there on whatever food or shelter Nature might provide, without seeking assistance from any human being.

I do not yet possess the spiritual strength
necessary for this. It seemed to me a base
thing, therefore, to accept food tainted by
war without working for it. When thousands
have come forward to lay down their lives only
because they thought it their duty to do so,
how could I sit still?

A rifle this hand will never fire. And so
there only remained nursing the wounded and
I took it up.[31]

In this highly significant letter, Gandhi voices the utter
helplessness, the hopelessness of the individual once the war
has started: there is absolutely no escape from the tentacles
that sprout from it. The extensive and ever-expanding interdependence of the individuals and groups makes avoiding
involvement with the war still more impossible. In all honesty,
therefore, if one perforce has to accept the daily bread
tainted in one way or the other by war, one must also then
work for it. And tending the wounded was, obviously, the
lesser evil than firing a rifle, which Gandhi, in any case,
was quite incapable of doing.

Gandhi, at this stage, did not seem sure as to whether
the steps he took--organizing the volunteer corps, appealing
and mobilizing the help of Indians in England, and so on--
were the right ones. But what he was absolutely clear about
was that he had done, and was doing, his duty in accordance with the teachings of the Gita.

The Gita says that he who eats without per-
forming Yajna is a thief. In the present situation here [in England] sacrifice meant, and
means self-sacrifice.

I saw, therefore, that I too must perform
Yajna. I myself could not shoot but I could
nurse the wounded. I might even get Germans
to nurse. I could nurse them without any
partisan spirit. There would be no violation
of the spirit of compassion then. And so I
decided to offer my services.

Now, I am not a private individual, but a
public figure. I must also talk to others. I
must address them an unconditional letter,
which I did.

My position . . . is on a level with the idea
that I must not kill a snake. But so long as,
in my cowardice, I fear a snake, I would certainly remove it to a distance, if not kill it
outright. This also is a form of violence.

If while I am removing one, it struggles
hard, I should hold it so tight between the
sticks that it might bleed, and even be
crushed to death. Even so, my statement that
I ought not to kill a snake would and must
hold true.

So long as I have not developed absolute
fearlessness, I cannot be a perfect Satyagrahi.[32]

Basically, it has to be remembered, Gandhi's participation in World War I was not much different in its nature
than that in two earlier occasions. It is true that on his
return to India, in response to the government's request, he
campaigned vigorously and sincerely--and in the face of
heavy odds precipitated by the noncooperation of his coun-
trymen--for enlistment and the military training in general.
Disappointment and at best a partial success did not deter
him from persevering. But he seems to have been quite
clear in his mind about two things: that the war should end
at the earliest time possible, and that the Indians should not
be asked to fight in the front lines, but, instead, "their
services will be utilised . . . for related purposes and for
looking after the welfare of the sepoys."[33] Evidently, the
Indians he seems to have had in mind were the ones he re-
cruited and not the regular soldiers who must, of course, do
their duty. His illness both in England and India perhaps
prevented him from going to the front. So his participation
this time--unlike the other two occasions--remained confined
almost entirely to the advocacy of military training for his
countrymen. Once again, his services to the Crown were
duly recognized by awarding him the Kaiser-i-Hind Gold
Medal well before the war actually came to an end.

ON WAR AFTER THE WAR

At the time of the Boer War, Gandhi's sympathies lay
with the Boers, but he had to keep his opinion to himself;
he could not impose his own feelings upon the state. His
lot was to owe allegiance and accord full support to the acts
of the state, and he advised his fellow Indians likewise. The
war may have been disastrous for others, but to him it
"came as a blessing in disguise" insofar as it offered him a
golden opportunity to prove the mettle of his countrymen,
whose miserable plight and callous exploitation in South
Africa he was agitating against.

Without doubt he had described war as antireligious and
said that as a Hindu he did not believe in it.[34] And yet his
firsthand experience at the front led him to claim that war

brings people closer to God. Addressing a meeting on January 27, 1902, in Calcutta, he candidly asked "how many proud rude savage spirits has [war] not broken [into] gentle creatures of God?" Thus he seems to have believed at the moment that justice lay on the side of the Boers and that something of value could after all emerge from the evil process of that war.

The value that seems to have emerged, we may venture, was curiously a somewhat romantic view of the British soldier. The monastic discipline in the military camps, the spirit of brotherhood among the soldiers, and their courage on the battlefield touched the romantic in Gandhi. He was quite stirred by the soldiers going to battle: they, he told himself, went not to take the life of others, but because it was their duty. He was to be cured of this romantic image in later years as he matured through an ever-deeper involvement with the issues arising during India's struggle for freedom and as he battled within himself in resolving the dilemma of individual action in an interdependent society. The mature Gandhi was then devoted to exploring the implications of his total rejection of war.

The Russo-Japanese War broke out in 1904 and further elicited Gandhi's views on the subject. Describing it as a war of giants,[35] he expressed his unhappiness and pain over the fact that thousands were being annihilated and mutilated by it; he fervently hoped for its early end. And yet his views seem to change as the war progresses. The editorials of the Indian Opinion started reflecting that Gandhi's sympathies were with the Japanese, and one finds him talking of virtues essential for war: "fervour is as necessary in other tasks as it is in war, and it is a positive virtue." He explained that Japan could defeat Russia at Port Arthur on August 10, 1904, because "she has been fighting with fervour." Further, "Japan did not receive any unexpected help, she had only the firm determination to win, and this determination has proved her true ally," and so on. Gandhi, at the time, not only admired the "epic heroism" of the Japanese but spelled out its rationale as well.

The secret of the Japanese victories in the war, according to Gandhi, lay in the sentiments of unity and burning patriotism and an unshakable resolve to do or die that animated the Japanese soldiers. Service to the nation, and the passionate belief that to render that service was the highest ideal, eliminated all kinds of differences and welded them into a powerful force. After all, one has to die one day; what does it matter if one dies on the battlefield? What is the certainty that staying at home and not going to the front will ensure longer life? And even if one did live very long, what good was it living as a subject people? Such, maintained Gandhi, was the reasoning that had made the Japanese defiant, contemptuous of death, and invincible on the battlefield.

But how would such a spirit, such idealism, and the flowing accounts of Japanese heroism avail the Indians? He exhorted the Indian community in South Africa then, and his countrymen at home later, to "emulate to some extent at least the example of Japan." At any rate, the line of reasoning he suggested for Japanese valor certainly found an echo during his campaign for recruits years later (1918) in India. En passant, one cannot help wondering whether his enthusiasm for Japan at the time did not reflect a degree of pride in the rise of an Asian power.

However, his eulogy of the Japanese spirit did not blind Gandhi to the evil that overtakes a society in war. Corruption eats like a canker into the vitals of such a society: taking advantage of the chaos, people commit misdeeds they would normally not think of; rules of honesty and good conduct are entirely forgotten; the values of jungle take over; and the bigger the war, the larger the scale of decay and confusion. He rightly pointed to the fraud and trickery prevalent during the Crimean and Boer Wars. And as a further illustration he cited the conditions of corruption and decay prevailing in Russia, which caused its humiliating defeat at the hands of Japan, even though Russia's gigantic Baltic fleet was equipped with the latest artillery.

Japan was superior, maintained Gandhi, also in that nobody there was exploiting the situation to personal advantage. In sum, he explained and justified the Japanese victory in the war on the grounds that Japan's demand was just, the society was free from corruption, individuals did their duty, and the people were united and lived in simplicity. In support of his assessment of the Russians in contrast, he cited Tolstoy.[36]

Gandhi's views on the Russo-Japanese War suggest two very significant points. First, in holding Japan's claim to be just, he seemed to believe that in war justice lay with one of the two sides; war could, therefore, enable justice to triumph. Second, he appreciated that security pacts and treaties between nations can have beneficial effects for the rest of the world, for this is the conclusion he drew from the Anglo-Japanese Treaty of January 30, 1902. He was radically to move away from the inference of the second of these points in his later years when he subscribed strongly to India's position of nonalignment, as will be discussed subsequently.

At the time of the Zulu uprising, Gandhi's focus was somewhat different. He sought equality of Indians with the other communities in South Africa by securing a permanent place for Indians in the ambulance corps constituted by the government and on quite different lines than done so far.

> That corps will be a permanent body; its members will be issued weapons, and they will re-

ceive military training every year at stated
times. For the present they will not have any
fighting to do. Wars are not fought all the
time.

. . . It is now more than 20 years since
the last Kaffir rebellion broke out. There is,
therefore absolutely no risk in joining the
Volunteer Corps. It can be looked upon as a
kind of annual picnic.

The person joining it gets enough exercise
and thus keeps his body in good trim and im-
proves his health. One who enlists as a volunteer is much respected. People love him, calling
him a civilian soldier. [Emphasis added][37]

Clearly, these remarks by Gandhi are addressed as much to
the Indians, as they were a part of his general plea for
equality to his countrymen. Otherwise, his stance was the
familiar one: admiration for the soldier for his discipline and
the spirit of comradeship (indeed, the training one gets in
discipline at the battlefield cannot be had anywhere else);
invoking Indians to make full use of the opportunity to prove
their worth; and so on.

Significantly, in pursuit of his objective--that of ensuring equality for the Indians--and in the flow of his language,
Gandhi, apparently gave a slight twist to his facts as, it
would seem to his conscience. It was simply not true that the
frontline soldiers lived on "health and happiness" or that
"fewer men died from bayonet or bullet wounds than from
sheer carelessness or perverse living" in the Crimean War.
Nor could he easily explain why he took the side of the British when he knew his sympathies lay with the Zulus. This
may seem to be a strange kind of "genuine sense of loyalty."
Except that Gandhi's participation at this time was more in
the spirit of a kind of satyagraha, as we have suggested
above, there seems to be no other plausible explanation.
However, the so-called rebellion did make Gandhi see the
real character of the British. Their naked and blatant atrocities against the Zulus pushed him to deplore war in general
and to call Britain "the Kingdom of Satan [the God of War]."[38]

From his experience of the "rebellion" on, conviction
started congealing in Gandhi that war was a degrading phenomenon, a brutalizing instrument of politics. In contrast,
self-suffering was a vastly superior weapon for resolving
conflicts. Thus Gandhi stated in his celebrated speech on the
Ethics of Passive Resistance at the Emerson Club, in London,
on October 8, 1909:

War with all its glorification of brute force is
essentially a degrading thing. It demoralizes

those who are trained for it. It brutalizes
men of naturally gentle character. It outrages
every beautiful canon of morality. Its path
of glory is foul with passions of lust, and red
with the blood of murder.39

By the time he encountered World War I, these strong sentiments of disapproval had further solidified to make all war
despicable in Gandhi's opinion. Personally, he would not kill
or injure anybody, friend or foe, of course. But more important, since war was contrary to the creed of a satyagrahi
--and Gandhi earnestly considered himself to be one--he
could not possibly "support war directly or indirectly."40
And yet, when war actually broke out in 1914, we find Gandhi voluntarily getting involved in the "degrading" process!
    Clearly, about this time, Gandhi was going through a
mental struggle; he had not matured, let alone perfected,
his creed of nonviolence. So his reaction to World War I was
somewhat mixed, if not seemingly confused. In his letter to
the pacifist A. H. West on November 20, 1914, he confessed:
"I share your views about the war. If I had the moral
strength, however, I would certainly be the passive resister
that you have pictured in your letter. Lately, I have come
to see the principle of non-violence in a somewhat different
light, sublime nonetheless."41 History, and the contemporary
situation, led Gandhi to think that

> War will always be with us. There seems to be
> no possibility of the whole human nature be-
> coming transformed. Moksha and Ahimsa [are]
> for individuals to attain. Full practice of ahimsa
> is inconsistent with the possession of wealth,
> land or rearing of children. There is real ahimsa
> in defending my wife and children even at the
> risk of striking down the wrongdoer. It is per-
> fect ahimsa not to strike him but intervene to
> receive his blows.
>   India did neither on the field of Plassey. We
> were a cowardly mob warring against one
> another, hungering for the Company's silver.
> . . . There was no ahimsa in their miserable
> performance, notwithstanding examples of per-
> sonal bravery. . . .42

Gandhi was surprised and pained that, among the difficulties he encountered during his recruitment campaign, fear
was essentially the most prominent: not a single person "objected [to being recruited] because he would not kill. They
object because they fear to die. This unnatural fear of death
is ruining the nation." This fundamental fact not only steeled

him to combat fear through recruitment, but at the same time
brought forth from him a very lucid rationale of his nonviolence in relation to India:

> India has lost the power to strike before she
> can voluntarily renounce the power to strike.
> She may never renounce. Then she will be as
> bad as the West or better still, the modernists.
> Today she is neither.
>
> The ancients in India knew the art of war-
> fare, the art of killing--and yet reduced the
> activity to a minimum and taught the world that
> it is better to refrain than to strike.
>
> Today, I find that everybody is desirous of
> killing but most afraid of doing so or powerless
> to do so. Whatever is to be the result I feel
> certain that the power must be restored to In-
> dia. The result may be carnage. Then India
> must go through it.[43]

Here, then, is the explanation of his demand for military
training and the art of modern warfare for Indians.

He was, however, firm in his belief that nonviolence was
still vastly superior as a weapon. Bravery, according to
Gandhi, lay in receiving the blows rather than in giving
them: "If our Government will fight with the Germans as it
does now, if our soldiers go and stand before them weapon-
less and will not use explosives and say 'We will die of your
blows', then, I am sure, our Government will win the war at
once."[44] Under the prevailing circumstances, Gandhi could,
of course, not advocate such a course for obvious reasons.
Moreover, the British could not offer this kind of resistance,
felt Gandhi, for it required samskar (acquired tendencies of
thought and character). India possessed the requisite samskar; the seeds of samskar could flourish in India, not in
England.

His esteemed and intimate friend, Dinabandhu C. F.
Andrews, had strong reservations about Gandhi's recruitment
campaign and objected that it was doing serious injury to the
cause of ahimsa. Indians, Andrews maintained, had always
repudiated blood lust and instead stood on the side of humanity. Gandhi did not accept these views[45] and held that, under the circumstances, not to campaign would be contrary to
his character:

> I have taken up [the recruitment campaign]
> to serve that very [ahimsa's] cause. I know
> that my responsibility is great. It was equally
> great when I was supine, feeling that recruit-
> ing was not my line. There was a danger of

those who put faith in my word becoming or
remaining utterly unmanly, falsely believing
that it was ahimsa. . . .
    One should have the ability in the fullest
measure to strike and then perceive the inability of brute force and renounce the power. . . .
    Jesus had the power to consume his enemies
to ask but he refrained and permitted himself to
be killed. . . .

Invoking the Gita, and the endless wars described in the
Mahabharta and Ramayana, Gandhi pointed out to Andrews
that Indians had always been warlike; "the finest hymn composed by Tulsidas in praise of Rama gives the first place to
his ability to strike down the enemy;" Buddhism had failed
precisely because it advocated forebearance; and during the
Mohammedan invasions, Indians were not any less eager than
their enemies to fight; under the British, arms had to be
renounced compulsorily but the fighting spirit had not entirely died down. Moreover, insisted Gandhi, he had never
asked Indians, "let us go and kill the Germans;" all he had
said was "let us go and die for the sake of India and the
Empire."
    The essential to be borne in mind here is that at this
time Gandhi never proclaimed that war was evil. Indeed, he
held "under exceptional circumstances war may have to be
resorted to as a necessary evil even as the body is;" "If the
motive is right it [war] may be turned to the profit of mankind and that an ahimsaist may not stand aside and look on
with indifference but must make his choice and actively cooperate or actively resist."[46] Moreover, his position that in
war the cause of one of the two sides can be just remained
unchanged up to this stage.[47] Gandhi also realized that
those "who go through the terrible strain of war without
collapsing must be Yogis," though "they would be fit for
moksha if their yoga was employed for a better cause."
Finally, then, he would still teach the use of force to those
who wanted to learn the art of fighting: the first principle of
ahimsa was that "all killing is not himsa."[48] Thus argued Gandhi in justification of his campaign for recruitment to the forces
of the empire.
    Interestingly, contrary to the prevailing viewpoint of
Western pacifists, Gandhi did not make any distinction between
ambulance work and combat at the front: "those who confine
themselves to attending to the wounded in battle cannot be absolved from the guilt of war."[49] His letter to Esther Faoring,
on June 30, 1918, while clarifying his position vividly reflects
his current state of mind:

    The choice before me was only that either we must
    renounce the benefits of the state or help it to the

>best of our ability. . . . Since Indians were not
>ready to renounce they must help Britain. . . .
>   I know I have put the argument most clumsily.
>I am passing through new experiences. I am
>struggling to express myself. Some of the things
>are still obscure to me. And I am trying to find
>words for others which are plain to me. I am
>praying for light and guidance and am acting with
>the greatest deliberation. . . .[50]

With this frame of mind, and faith in God, Gandhi soldiered on, following "His Will and no other" in the conviction that "He will lead me amid the encircling gloom," until the war came to an end.

NOTES

1. Joseph J. Doke, M. K. Gandhi, An Indian Patriot in South Africa (Varanasi: Akhil Bharat Sarv Seva Sangh Prakashan, 1959).

2. The Collected Works of Mahatma Gandhi, vol. 3 (New Delhi: Government of India, Publications Division), pp. 113-114, 119-121, 126-127, 129-130, 216.

3. Ibid., pp. 138-139; also pp. 133fn, 137, 222.

4. Ibid., pp. 164, 216.

5. Ibid., vol. 5, pp. 233-234, 291, 311-312, 366-367.

6. Ibid., pp. 348, 357.

7. Ibid., p. 372.

8. Ibid., vol. 3, pp. 376-377.

9. Ibid., vol. 12, p. 523.

10. Ibid., p. 527.

11. The letter was signed by Gandhi, his wife Kasturba, Sarojini Naidu, and over 50 Indian doctors, lawyers, and students living in London. Ibid., p. 528.

12. Ibid., pp. 642-643.

13. Mahatma Gandhi, Hind Swaraj (Ahmedabad: Navajivan, 1962), pp. 58-60.

14. Collected Works, vol. 12, pp. 538-539, 542.

15. "I have had to start a Satyagraha here against the India Office; the Satyagraha is over. We got what we wanted." Ibid., pp. 546-547.

16. Ibid., pp. 555, 565.

17. Ibid., vol. 14, p. 371 n 3.

18. On his way to the conference, Gandhi had asked Charles Frere Andrews--a close associate of his, who did not approve of Gandhi associating himself with war (ibid., p. 444 n 3)--to join him, which the latter did. In the train, Andrews read in the New Statesman (London) an account of the predatory "secret treaties" supposedly unearthed by the revolutionaries from the Russian Foreign Office; Great Britain was a signatory of these treaties. Andrews thrust the account before Gandhi, demanding, "How can you take part in a war conference while this sort of double dealing is going on? Ibid., p. 372.

19. Ibid., p. 375. In his Autobiography, Gandhi wrote, "So I attended the conference. The Viceroy was very keen on my supporting the resolution about recruiting. I asked for permission to speak in Hindu-Hindustani. The viceroy acceded to my request, but suggest that I should speak also in English. I had no speech to make. I spoke but one sentence to this effect: 'With a full sense of my responsibility, I beg to support the resolution.'" Mahatma Gandhi, An Autobiography, My Experiments with Truth (Ahmedabad: Navajivan, 1959), p. 327.

20. Collected Works, vol. 14, pp. 377-382, 402, 479.

21. Ibid., p. 410.

22. Ibid., pp. 422-424, 433, 435-438, 483; vol. 15, pp. 1, 15.

23. Robert Payne, The Life and Death of Mahatma Gandhi (London: Bodley Head, 1969), p. 120.

24. Cited in C.F. Andrews, Mahatma Gandhi's Ideas (London: Allen & Unwin, 1949), p. 364.

25. Collected Works, vol. 3, p. 160; also pp. 135 including n 4, 133-134, 136, 137-141.

26. Ibid., vol. 5, pp. 234, 362.

27. Ibid., vol. 3, pp. 146-148, 159-162, 216-217, 219-224.

28. Ibid., vol. 5, p. 134, 154.

29. L.S. Amery, ed., Times History of War in South Africa, vol. 3 (London: Sampson, Low, Marsten, 1905).

30. D.G. Tendulkar, Mahatma, vol. 1 (New Delhi: Government of India, Publications Division, 1960), p. 76.

31. Collected Works, vol. 12, pp. 531-532; vol. 5, p. 353.

32. Ibid., pp. 554-555.

33. Ibid., vol. 14, p. 443.

34. Ibid., vol. 3, pp. 216, 223.

35. Ibid., vol. 4, p. 374; also, pp. 431, 466-467.

36. "Tolstoy is still writing with great energy. Though himself a Russian, he has written many strong and bitter things against Russia concerning the Russo-Japanese War. He has addressed a bitter and pungent letter to the Czar in regard to the War." Ibid., vol. 5, p. 57; also, pp. 30-31, 41.

37. Ibid., p. 362; also, pp. 353, 366.

38. Ibid., vol. 10, p. 189.

39. Ibid., vol. 9, p. 471; vol. 14, p. 380.

40. Ibid., vol. 12, p. 554.

41. Ibid., pp. 556-557; vol. 14, p. 448; vol. 15, p. 19.

42. Ibid., vol. 14, pp. 509-510.

43. Ibid., p. 520.

44. Ibid., pp. 407-408.

45. Ibid., p. 511; also, p. 475.

46. Ibid., p. 477.

47. Letter of April 26, 1918 to Sir Claud Hill. Ibid., p. 372; also, p. 403.

48. Ibid., p. 485.

49. Ibid., vol. 12, pp. 531, 555; also, Autobiography, p. 258.

50. Collected Works, vol. 14, pp. 462-463; vol. 15, p. 4.

3
PARTICIPATION RATIONALE

To a world used to remembering Gandhi as an apostle of nonviolence, it may appear to be something of an enigma that he not merely participated in war after a fashion, but even justified his action--neither deviating much from his chosen position nor, indeed, regretting his participation even later on. His own contemporaries were quite mystified by his behavior.

Dinabandhu C. F. Andrews, so close to Gandhi and such an admirer of his, found himself unable to accept Gandhi's position.[1] The Home Rule Leaguers, also differing sharply, did not cooperate with him during World War I. In a running debate during the years after the war, the pacifists in Europe and the United States assailed him, demanding explanation for his involvements in war to the detriment of his own creed of nonviolence.

Let the case for the "prosecution" be stated first, and we can do no better than cite a correspondent in search of an answer to the riddle of Gandhi's seemingly contradictory posture:

> When the Zulus broke out for liberty against the British usurpers, you helped the British in suppressing the so called rebellion.
> Is it a rebellion to try to shake off the foreign yoke? Was Joan of Arc a rebel? Was George Washington a rebel? Is De Valera one?
> You may say that the Zulus had recourse to violence. I then ask, was the end bad or the means? The latter may have been so, but certainly not the former. . . .
> In the last War, when the gallant Germans and Austrians were fighting against a world

combination, you raised recruits for the British to fight against the nations that had done
India no harm.

Whenever there is a war between two races,
one has to hear both parties before coming to
a decision, either for or against any of them.
. . .

You have all along been an advocate of
passive resistance and non-violence. Why then
did you induce people to take part in a war
the merits of which they knew not, and for the
aggrandizement of a race so miserably wallow-
ing in the mire of imperialism. . . .[2]

In response, we will let Gandhi state his position in his
own words:

Not only did I offer my services at the time
of the Zulu Revolt but before that at the time
of the Boer War, and not only did I raise
recruits in India during the late War, but I
raised an ambulance corps in 1914 in London.
If therefore I have sinned, the cup of my sins
is full to the brim. I lost no occasion of serving the Government at all times.

Two questions presented themselves to me
during all those crises. What was my duty as
a citizen of the empire as I then believed myself to be, and what was my duty as an out
and out believer in the religion of ahimsa--non-
violence?

I know now, that I was wrong in thinking
that I was a citizen of the empire. But on
those four occasions I did honestly believe that
in spite of the many disabilities that my country was labouring under, it was making its way
towards freedom, and that on the whole the
Government from the popular standpoint was
not wholly bad and that the British adminis-
trators were honest though insular and dense.

Holding that view, I set about doing what
an ordinary Englishman would do in the circumstances. I was not wise or important enough
to take independent action. I had no business
to judge or scrutinize ministerial decisions with
the solemnity of a tribunal. I did not impute
malice to the ministers either at the time of the
Boer War, the Zulu Revolt or the late War. I
did not consider Englishmen, nor do I now consider them as particularly bad or worse than

other human beings. I considered and still
consider them to be as capable of high motives
and actions as any other body of men and
equally capable of making mistakes. I therefore
felt that I sufficiently discharged my duty as
a man and a citizen by offering my humble
services to the empire in the hour of its need
whether local or general. That is how I would
expect every Indian to act by his country
under Swaraj.

The next point, that of ahimsa, is more
abstruse.

My conception of ahimsa impels me always
to dissociate myself from almost every one of
the activities I am engaged in. My soul refuses
to be satisfied so long as it is a helpless witness of a single wrong or a single misery.

But it is not possible for me, a weak, frail,
miserable being to mend every wrong or to
hold myself free of blame for all the wrong I
see. The spirit in me pulls one way, the flesh
in me pulls in the opposite direction. There is
freedom from the action of these two forces, but
that freedom is attainable only by slow and painful stages. I cannot attain freedom by a mechanical refusal to act, but only by intelligent action
in a detached manner. The struggle resolves
itself into an incessant crucifixion of the flesh
so that the spirit may become entirely free.
[Emphasis added]

"I was again an ordinary citizen," continued Gandhi,

no wiser than my fellows, myself believing in
ahimsa and the rest not believing in it at all
but refusing to do their duty of assisting the
Government because they were actuated by
anger and malice. They were refusing out of
their ignorance and weakness.

As a fellow-worker it became my duty to
guide them aright. I therefore placed before
them their clear duty, explained the doctrine
of ahimsa to them and let them make their
choice which they did. I do not repent of my
action in terms of ahimsa. For under swaraj too
I would not hesitate to advise those who would
bear arms to do so and fight for the country.
[Emphasis added] 3

He offered his services not because he believed in war, but simply because he "could not avoid participation in it at least indirectly." He could not resist it even if he had wanted to, for what was his status? War cannot be avoided by not taking part in it anymore than evil can be avoided by not participating in it. "This," emphasized Gandhi, "needs to be distinguished from sincerely helpless participation in many things we hold to be evil or undesirable."4

To the charge that enlisting men for field service amounted to helping the cause of war and thus did not, obviously, tally with his own principle of ahimsa, Gandhi's reply was:

> By enlisting men for ambulance work in South
> Africa and in England, and recruits for field
> service in India, I helped not the cause of war,
> but I helped the institution called the British
> Empire in whose ultimate beneficial character I
> then believed.
>
> My repugnance to war was as strong then
> as it is today and I could not then have, and
> would not have shouldered a rifle.
>
> But one's life is not a single straight line;
> it is a bundle of duties often conflicting. And
> one is called upon continually to make one's
> choice between one duty and another as a citi-
> zen. . . .5

As to what impelled him to participate in war, Gandhi confessed to having done so with a mixture of motives:

> Two things I can recall. Though as an individ-
> ual I was opposed to war, I had no status for
> offering effective non-violent resistance.
>
> Non-violent resistance can only follow some
> real disinterested service, some heart-expres-
> sion of love. For instance, I would have no
> status to resist a savage offering animal sacri-
> fice until he could recognise in me his friend
> through some loving act of mine or other means.
>
> I do not sit in judgement upon the world
> for its many misdeeds. Being imperfect myself
> and needing toleration and charity, I tolerate
> the world's imperfections till I find or create an
> opportunity for fruitful expostulation.
>
> I felt that, if by efficient service I could
> attain the power and the confidence to resist
> the Empire's wars and its warlike preparations,
> it would be a good thing for me who was seek-
> ing to enforce non-violence in my own life, to

test the extent to which it was possible among
the masses.

The other motive was to qualify for Swaraj
through the good offices of the statesmen of the
Empire. I could not thus qualify myself except
through serving the Empire in its life and death
struggle.6

However, did such a motive not violate his interpretation
of the Gita, which exhorts us to act but never with a view
to the fruits of our action? Gandhi maintained that there was
no contradiction, for "the verse referred to . . . has a double meaning: one is that there should be no selfish purpose
behind our actions. That of gaining Swaraj is not a selfish
purpose. Secondly, to be detached from fruits of action is
not to be ignorant of them, or to disregard or disown them.
To be detached is never to abandon action because the contemplated result may not follow. On the contrary, it is proof
of immovable faith in the certainty of the contemplated result
following in due course."7

Another thrust of the attack came from the Belgian pacifist Barthélemy de Ligt, who in an open letter took Gandhi
to task for having participated in the wars of the British Empire and called upon him to explain himself in the light of
his proclaimed faith in nonviolence.8 Gandhi used the columns
of Young India of 1928-29 to elucidate his position:

There is no defence for my conduct weighed
only in the scales of Ahimsa. I draw no distinction between those who wield the weapons of
destruction and those who do Red Cross work.
Both participate in war and advance its cause.
Both are guilty of the crime of war.

But even after introspection during all these
years, I feel that in the circumstances in which
I found myself, I was bound to adopt the course I
did both during the Boer War and the Great European War, and for that matter the so-called
Zulu Rebellion of Natal in 1906.

Life is governed by a multitude of forces.
It would be smooth sailing if one could determine
the course of one's actions only by one general
principle whose application at a given moment
was too obvious to need even a moment's reflection. But I cannot recall a single act which
could be so easily determined.

Being a confirmed war resister I have never
given myself training in the use of destructive
weapons in spite of opportunities to take such
training. . . .

It was perhaps thus that I escaped direct destruction of human life. But so long as I lived under a system of Government based on force and voluntarily partook of the many facilities and privileges it created for me, I was bound to help that Government to the extent of my ability when it was engaged in a war, unless I non-co-operated with that Government and renounced to the utmost of my capacity the privileges it offered me.

Even so did I participate in the three acts of war. I could not, it would be madness for me to, sever my connections with the society to which I belong. And on those three occasions I had no thought of non-co-operating with the British Government.[9]

With regard to the alleged infringement of his faith in non-violence, Gandhi stated that it

works in a most mysterious manner. Often a man's actions defy analysis in terms of non-violence; equally often his actions may wear the appearance of violence when he is absolutely non-violent in the highest sense of the term and is subsequently found to be so.

All I can then claim for my conduct is that it was in the instances cited, actuated in the interest of non-violence. There was no thought of sordid national or other interest. I do not believe in the promotion of national or other interests at the sacrifice of some other interest.[10]

The pacifists, however, would not be pacified with this defense. They continued to feel disturbed over Gandhi's attitude. So, to pacifist Vladimir Tcherkoff, Gandhi had to write categorically:

It is [a matter of] deep confiction that war is an unmixed evil. I would not yield to anyone in my detestation of war.

But conviction is one thing, correct practice is another. The very thing that one war resister may do in the interest of his mission may repel another war resister who may do the exact opposite, and yet both may hold the same view about war.

This contradiction arises because of the bewildering complexity of human nature. I can only

> therefore plead for mutual toleration even
> among professors of the same creed.11

But this should <u>not</u>, insisted Gandhi, be taken even remotely
as an expression of any kind of regret: he was not sorry
that he had aided the British. Nor had his participation in
the said wars been out of expediency. "I claim to have done,"
said Gandhi, "every act described by me for the purpose of
advancing the cause of peace. That does not mean that those
acts really advanced the cause of peace. I am merely stating
the fact that my motive was peace." His way of looking at
peace, however, was very different from that of the European pacificists. He looked at peace, he pointed out, "through
a medium to which my European friends are strangers. I belong to a country which is compulsorily disarmed and has
been held under subjection for centuries. My way of looking
at peace may be necessarily different from theirs."12 Belonging thus to a people without any choice or status, Gandhi
said that till "we come to our own, we shall have to be victims of the war that may come upon the world." This attitude
was reflected during World War II, when he allowed the Allied
troops to be based in and operate from the soil of India,
while he himself refused firmly to be associated in any way
with the war of the empire.

Gandhi took the opportunity to point to a fallacy in the
stand of the Western pacifists: "Fellow-resisters in the west
are participants in war even in peace time in as much as they
pay for the preparations that are being made for it and otherwise sustain governments whose main occupation is such
preparation."13

In summation we cite the following passage by Gandhi:

> It is profitless to speculate whether Tolstoy in
> my place would have acted differently from me.
> It is enough for me to give the assurance to my
> friends in Europe that in no single act of mine
> have I been consciously guilty of endorsing vio-
> lence or compromising my creed.
>
> Even the seeming endorsement of violent
> action by my participation on the side of Britain
> in the Boer War and Zulu Revolt was a recogni-
> tion in the interest of non-violence, of an inevi-
> table situation.
>
> That the participation may nevertheless have
> been due to my weakness or ignorance of the
> working of the universal law of non-violence is
> quite possible. Only I had no conviction then,
> nor have any now, of such weakness or igno-
> rance.14

It is clear, then, that Gandhi fully justified his participation not only during the course of the wars, but also in
the 1920s, reiterating his earlier arguments, which he saw
no reason to revise materially.[15] Nor did he seek to mitigate
his involvement and its implications by pleading the humanitarian nature and noncombatant status of his participation.
Indeed, he held himself equally "guilty of the crime of war."[16]
He conceded that his involvement might have been against the
principle of nonviolence, but there was perfect justification
for it. Even during those years he recognized "the immoral-
ity of war"[17] but, at the same time, was fully clear and convinced about the rightness of his action. Being "a votary of
truth," he was "obliged to grope in the dark."[18] The truth
for him lay in doing his duty as a citizen of the British Empire. So Gandhi read the situation entirely in terms of duty
rather than anything else. It must be stressed that at no time
in his life had Gandhi ever said that he approved of, or believed in, war as an institution for resolving interstate disputes or as an instrument of national policy. Indeed, he emphatically and repeatedly decried war. "My opposition to and
disbelief in war," he steadfastly maintained,

> was as strong then as it is today. But we have
> to recognise that there are many things in the
> world which we do although we may be against
> doing them. I am as much opposed to taking the
> life of the lowest creature alive as I am to war.
> . . .
> Possession of a body like every other pos-
> session necessitates some violence. . . . The
> fact is that the path of duty is always easy to
> discern amidst claims seeming to conflict with
> the other. . . .[19]

Apparently, Gandhi recognized, and accepted, that some
violence in life was unavoidable: as long as human beings
remain in flesh they simply cannot avoid acts of violence.

Manifestly, the whole confusion about the implications
of Gandhi's participation in war arises from a totally errone-
ous assumption, namely, that Gandhi was a pacifist. Bart de
Ligt persisted in his accusation that Gandhi sacrificed his
pacifism in order to win the heart of the English.[20] Various
other Western pacifists and intellectuals first branded Gandhi
arbitrarily with this quite unwarranted stamp of pacifism and
then proceeded to pile their scorn and unsparing censure
upon him.

Thus, the International Encyclopaedia of Social Sciences
classifies Gandhi among the pacifists.[21] Mulford Sibley, an
exponent of pacifism, described Gandhi as perhaps the greatest pacifist of the twentieth century "who combined in his

views and practices both religious and utilitarian notes."
Pacifist-religious and humanitarians throughout the world
claimed Gandhi to be one of them. Interwar years, which,
incidentally, marked the heyday of pacifism, had the pacifists seeing in Gandhi an unflinching follower of the Sermon
on the Mount for he refused to meet evil with evil or violence with violence, declared all war to be evil, and offered
a noble technique of combating evil. This they inferred from
the fact that Gandhi lived in the mood of scriptual texts:
the sermon exhorts "resist not evil" and teaches that retaliation does not lead to the desired end. Gandhi not only
preached that; he passionately believed in the brotherhood
of humanity and in the Christian doctrine of love. Gandhi,
to pacifists, was thus a unique living example of what they
had dreamed and what they were in pursuit of. They, accordingly, proclaimed him to be a "pacifist."[22] And Tolstoy's
influence on him, such as it was, had of course been interpreted in pacifist terms.

Perhaps the sharpest, and most patently misplaced, in-
dictment that Gandhi received was at the hands of the U.S.
theologian and political preacher, Reinhold Niebuhr (1893-
1971). He was at one time a thoroughgoing Christian Socialist
who evoked the biblical prophets and a bit of Marx in his
vituperous campaign against exploitation of labor. Niebuhr
had supported the U.S. entry into World War I against Germany. Having been raised in a German-speaking home, he at
that time was struggling to master English. He ran for Congress on a Socialist ticket but duly came to accept capitalism.
Niebuhr led the pacifist Fellowship of Reconciliation but
railed against U.S. isolationism in world politics. He supported once again the U.S. entry into World War II. Thoroughly distrustful of communism, Niebuhr became a kind of
intellectual chaplain of the cold war. According to him, the
atomic weapons are "our ultimate insecurity and our immediate security." He considered himself a "teacher of social
ethics" rather than a theologian.

This was the man, the apostle of Christian realism, who,
following his disillusionment with pacifism about the time of
the Great Depression declared that pacifism was no answer
to the struggle in the world to attain social justice, and that
violent coercion--not an evil in itself--was necessary.[23] The
rise of fascist-nazi dictatorships and the concomitant mood of
Europe had had its impact on the pastor. He had no patience
with pacifism. In the chagrin of his vitriolic attacks on pacifism, whenever Niebuhr thought of pacifists, he thought of
Gandhi whom he promptly made virtually the sole target of
his ire. He charged that among other things Gandhi's ethics
of love was divorced from justice and social responsibility.[24]
This speaks enough about the objectivity of Niebuhr and his
understanding of Gandhi.

Krishanlal Shridharani, on the other hand, while not denying that Gandhi was a pacifist emphasizes that he was not merely so.[25] He makes a subtle distinction between pacifism and Gandhi's stand and points out that, when the pacifists faced the dilemma of either succumbing to the militarists or joining the ranks of inactive pious wishers, they found a way out in Gandhi's concept of satyagraha.[26] Sridharani, thus, willy-nilly assigned Gandhi to the domain of pacifism.

Reginald Reynold, the celebrated Quaker campaigner for peace and a great friend of India, seems to depict Gandhi as the savior of pacifists.[27] Another writer on Gandhi, Paul F. Power, satisfied that Gandhi was a pacifist, makes the point that Gandhi's pacifist posture varied in accordance with the circumstances, especially in proportion to the demands of his nationalism. "Rather vexingly," holds Power, "Gandhi's ideas about war cut across unqualified pacifism, conditional pacifism and patriotic realism."[28] These ideas, he elaborates, developed almost chronologically following Gandhi's association with war.

To fully appreciate Gandhi's position with regard to war, it is very essential to be quite clear about the precise meaning and connotation of pacifism and the view Gandhi took of those connotations. Gene Sharp defines pacifism as "a belief system of those who, as a minimum, refuse participation in all wars on moral, ethical or religious principles."[29] Arising as a reaction to the widespread death, destitution, and destruction caused by war, pacifism became, essentially, an extremist antiwar movement demanding outlawry of individual or collective participation in war under any circumstances. The philosophy adopted by the pacifists has a long history going well beyond the New Testament to the seventh or sixth centuries before Christ.[30] As a modern concept, pacifism became fashionable about the time of World War I, when anything and everything that questioned the inevitability or desirability of war or sought to prevent war and promote peace came to be designated as pacifism. In the interwar period the rise of fascism, nazism, and totalitarian regimes in Europe imparted to pacifism a rather specific connotation: namely, that of a cult that aims at avoiding, or suppressing, war by urging resistance through political and social policies of the various countries. The cult steadily gained currency in Britain and the dominions, France, and the United States among others. A typical expression of this cult was the well-known motion in the 1933 Oxford Union debate that "This House will in no circumstances fight the wars for its King and Country," carried by 750 to 138. Notable among the groups subscribing to and propagating these attitudes of pacifism were the conscientious objectors and the Quakers, who manifested their bitter opposition to war by condemning conscription and refusing to pay taxes for military purposes.

Evidently, then, the starting point of such pacifists is their refusal to participate in war because it is contrary to the law of life. Their basic motive is negative, namely, the avoidance of war. Their stand is as simple, and cut and dried, as that. It is not for them to pause and take into account a host of other considerations inevitably and inextricably entailed in every situation that, in turn is inexorably dynamic. They would not budge in the slightest from the rather elementary position of not associating with war under any condition. For such pacifists, it would thus seem, war need not be related to justice.

Gandhi, however, does not fit into such a one-dimensional and arbitrary framework for a variety of reasons. In the first place, Gandhi's attitude toward war--unlike the pacifists who just adopted it--was steadily developed over the years in a series of experiments with truth. He grew to his attitude, while the pacifists picked it up from one source or the other because it was there. Secondly, Gandhi's basic problem was not war per se. His primary, and intense, concern was the question of human dignity in all its aspects. As a result, he unceasingly labors the aspect of duty as a citizen: for his countrymen he prescribes performance of duty as a sure passport to equality, as a release from the stranglehold of contempt and exploitation by the whites. In contrast to that of the pacifists, Gandhi's starting point is, accordingly, very different: it is not opposition to war. Hence, he called to the Western pacifists to appreciate one very vital difference between their situation and his own:

> They [Western pacifists] do not represent exploited nations; I represent the most exploited nation on earth.
>
> To use an unflattering comparison, they represent the cat and I represent the mouse. Has a mouse even the sense of non-violence? Is it not a fundamental want with him to strive to offer successful violence before he can be taught to appreciate the virtue, the grandeur, the supremacy, of the law of non-violence-- Ahimsa--in the field of war?
>
> May it not be necessary for me, as a representative of the mouse tribe, to participate in my principal's desire for wreaking destruction even for the purpose of teaching him the superiority of non-destruction?[31]

Without the slightest doubt, personally, Gandhi was absolutely uncompromising in his opposition to all war; nor did he want anybody to use his participation in certain wars to justify war under any conditions.[32] "I know," he said, "that war is wrong, is an unmitigated evil. I know too that it has to go. . . .

> Would that all the acts alleged against me were
> found to be wholly indefensible rather than
> that by any act of mine non-violence was held
> to be compromised, or that I was ever thought
> to be in favour of violence or untruth in any
> shape or form.
> Not violence, not untruth but non-violence
> and truth is the law of our being.[33]

At the same time, he asserted, that in war he was quite willing to take sides and to hold that one of the two parties in combat was in the wrong. His unhesitating and prompt moral support to the democracies against the Axis powers in World War II amply illustrates this:

> I believe all war to be wholly wrong. But if we
> scrutinise the motives of the two warring par-
> ties, we may find one to be in the right and the
> other in the wrong. For instance, if A wishes
> to seize B's country, B is obviously the wronged
> one. . . .
> I do not believe in violent warfare, but all
> the same B whose cause is just, deserves my
> moral help and blessings.[34]

In comparison with this clearly ennunciated position, one simply does not find any pacifist relating war to justice and/or holding that in a conflict the cause of one of two parties is just.

More significant is the fact that Gandhi is not content with merely deciding that one of the two warring sides has justice on its side. Indeed, he calls upon the pacifist to "wish success to the one who has justice on his side. By so judging he is more likely to bring peace between the two than by remaining a mere spectator."[35]

So whereas pacifists held that war was negation of justice in all respects and must be abjured for the simple and all-embracing reason that it was war, Gandhi, in sharp contrast, found war quite justifiable in certain circumstances. For instance, in the context of the Nazi persecution of Jews, Gandhi categorically stated, "If there ever could be a justifiable war in the name of and for humanity, a war against Germany, to prevent the wanton persecution of a whole race, would be completely justified."[36] He declared this at a time when he had irrevocably decided to stay away and condemn war unequivocally. But the mahatma in him could not remain callously indifferent to the brutal reality of life and the moral issues at stake in World War II; mass-scale murder had to be checked.

Gandhi, in fact, had no use for pacifism as an ideology of protest against violence on mere grounds of conscience. Contrary to the stolid passivity of the pacifist, his was always an ardent and constant participation in an area of conflict, and that is precisely what he demanded of the pacifists. Pacifism, as a moral posture, just did not appeal to Gandhi:

> To refuse to render military service when the particular time arrives is to do the thing after all the time for combating the evil is practically gone.
>
> Refusal of military service is much more superficial than non-co-operation with the whole system which supports the state.37 [Emphasis added]

The subtle point that Gandhi is making here is significant. He contended that it is very easy to say that war is anti-religious or immoral. Such an attitude, however, would evidently amount to hypocrisy unless the person concerned had taken that position, and actively defended it, before the commencement of hostilities. It would not be very moral to assume this stance to justify abstention after the hostilities had broken out.

We emphasize once again the importance that Gandhi placed on action, which is inexorably and inextricably linked with reality. He maintained that it was a sin against ahimsa for a person to sit passively until "the existing rule of spoliation had ended."38 "A non-violent man," he maintained,

> will instinctively prefer direct participation to indirect, in a system which is based on violence and to which he has to belong without any choice being left to him.
>
> I belong to a world which is partly based on violence. If I have a choice only between paying for the army of soldiers to kill my neighbours or to be a soldier myself, I would, as I must, consistently with my creed, enlist as a soldier in the hope of controlling the forces of violence and even of converting my comrades.39

Instead of defending himself, Gandhi took the position that all human existence is built upon evil and that our choice is limited to taking the lesser evil rather than to the avoidance of evil altogether. The law of karma, by enjoining upon us the sheer desirability of action, seeks to prevent the growth of fatalistic attitudes. Accordingly,

> When two nations are fighting the duty of a
> votary of ahimsa is to stop the war.
> He who is not equal to that duty, he who
> has no power of resisting war, he who is not
> qualified to resist war, may take part in war,
> and yet whole-heartedly try to free himself,
> his nation and the world from war.[40]

It is instructive to mention here that Gandhi, the great
innovator, who had applied satyagraha on an unprecedented
scale, cautioned Indians against taking the law into their
hands against their own government in times of national crisis. Referring to his participation in war, Gandhi acknowledged that he was satisfied that his help to the empire in its
time of need had enabled him to discharge his duty as a citizen and as a man. And this is what he expected his countrymen to do in a free India.[41] As shall be seen in the next
chapter, this is precisely what he did in the October 1947
rape of Kashmir.

Gandhi's attitude toward military training also was far
more flexible than that of the pacifists or war resisters. His
life's endeavor had been to create a nonviolent India where
there would be only an enlightened police to maintain law and
order.[42] Yet he was realistic enough to appreciate that a
free India may well have to maintain a military establishment:

> If there is a national government, whilst I should
> not take any direct part in any war, I can con-
> ceive of occasions when it would be my duty to
> vote for the military training of those who wish
> to take it. For I know that all its members do not
> believe in non-violence to the extent I do.
> It is not possible to make a person or a soci-
> ety non-violent by compulsion.[43] [Emphasis
> added]

In another passage, Gandhi explains, "Whilst, therefore,
even as an out and out believer in ahimsa, I can understand
and appreciate military training for those who believe in the
necessity of the use of arms on given occasions."[44]

In the light of the discussion above, we find that the
pacifists' assumption of Gandhi as being one of them and
subsequent censure of him for not measuring up to their
concept of pacifism becomes untenable. Gandhi's faith in non-
violence included, and accorded, due place to the concepts
of justice and responsibility. It is fear that has no place in
Gandhi's scheme of things and in nonviolence: "The word
fear can have no place in the dictionary of ahimsa,"[45] he
asserted repeatedly.

Accordingly, Gandhi's starting point was freedom from fear rather than war. The Pietermauritzburg experience gave Gandhi his initial release from fear while, simultaneously, making him conscious that submitting to injustice out of fear is cowardly; to slink away is still more dastardly. Injustice must be resisted and fought, and for doing that all fear must be cast out. His entire life is a saga of relentless striving for becoming fearless.

Gandhi found his countrymen living in the grip of all-pervasive oppressive and strangling, stunting fear: fear of police, of the officials, of suppressive laws, of prisons, fear of the money lender and the landlord--indeed, an inexhaustible, interminable variety of fear arising from a myriad of sources. Unless they came out of this damning, dominant impulse of fear, felt Gandhi, there could be no salvation. As long as fear had its grip, they could not possibly rise against injustice. So Gandhi urged his countrymen to be fearless, unbending, and free to be able to overcome injustice in all its facets.

The black pall of fear could not be destroyed in a trice, of course; ways and means must be found and/or forged. Nonviolence was the method Gandhi evolved for that purpose and assured his countrymen that it was the most superior technique of fighting tyranny and injustice. Evidently, freedom from fear became the first prerequisite of his technique of nonviolence, in that as long as fear existed one could not adopt nonviolence: "he who has not overcome all fear cannot practice ahimsa to perfection."46 The very decision to become nonviolent signified the conquest of fear.

Fear, moreover, was the most irrational element in human nature: "fear is a thing which I dislike; why should one man be afraid of another man" Gandhi asked.47 Fear makes people cowards and renders them helpless. Thoroughly a man of action, a karmayogi in fact, Gandhi could not tolerate helplessness and cowardice: "It is only because we have created a vicious atmosphere of impotence round ourselves that we consider ourselves to be helpless even for the simplest possible things."48

Accordingly, Gandhi wanted people to break this strangling vice of fear, come out of the state of helplessness, and fight injustice with the supreme weapon of nonviolence. His ahimsa, thus, inevitably includes truth and fearlessness; he is convinced that fearlessness is "indispensable for the growth of other noble qualities." He advises that in times of national crises we have two choices: we can be a military power, or, if we follow the Gandhian path, we "can become a great nonviolent and invincible power. In either case the first condition is shedding of all fear."49

It was with the intention of inculcating fearlessness in his countrymen--besides, of course, the objectives of winning the status of equality and self-respect for them--that made Gandhi advise them to volunteer during war:

> As a citizen not then [during World War I], and not even now, but as a reformer leading an agitation against the institution of war, I had to advise and lead men who believed in war but who, from cowardice or from base motives or from anger against the British Government refrained from enlisting.
>
> I did not hesitate to advise them that so long as they believed in war and professed loyalty to the British constitution, they were in duty bound to support it by enlistment.

Therefore, according to Gandhi, it was

> better [to] commit violence than sit helplessly out of cowardice in the name of non-violence.
> . . .
> To the coward I shall not be able to deliver my message of non-violence. Him I shall not be able to teach the lesson of peacefulness; I will be able to give the lesson of peace, the lesson of non-violence, only to those who do not fear to die.[50]

In Gandhi's thought and action, consequently, there is an unsparing and constant emphasis on fearlessness, because in his considered opinion, fear is the greatest evil that can befall a nation and must, therefore, be taken care of first; all other evils are secondary and could be attended to later. It is in this context that, even when he holds nonviolence to be infinitely superior to violence, he sanctions and condones violence in certain circumstances: while "all violence is bad and must be condemned in the abstract,"[51] Gandhi does not wish that you should "eschew violence in your dealings with robbers or thieves or with nations that may invade India." Thus, when it is a question of the survival of a nation, for instance, he would "risk violence a thousand times than risk the emasculation of a whole race."[52] Similarly, violence "when . . . offered in self-defence or for the defence of the defenceless, it is an act of bravery far better than cowardly submission."[53] Proclaiming that "cowardice is impotence worse than violence," Gandhi would always prefer resorting to arms rather than succumbing to cowardice. Were India's choice to be limited either to a recourse to arms or to cowardice, Gandhi would unhesitatingly want India to exercise the former option and defend its honor rather than be a passive, helpless witness to its own dishonor through cowardice.[54] Indeed, in his undiluted abhorrence of cowardice, Gandhi is even willing to revise his opinion with regard to revolutionary activity: "Cowardice, whether philosophical or otherwise, I abhor. And if I could be persuaded that revolutionary activity has dispelled cowardice, it will go a long way to soften my abhorrence of the method however much I may still oppose it on principle."[55]

Further, Gandhi firmly believes that nonviolence can only be used by the strong: "The doctrine of nonviolence is not for the weak and the cowardly; it is meant for the brave and the strong. The bravest man allows himself to be killed without killing. And he desists from killing or injuring because he knows that it is wrong to injure. . . ."[56] Obviously, courage is another virtue Gandhi values. It was the courage of the soldiers and warriors that impressed him tremendously and made him speak so highly of them. And when he talked of some good coming out of war, besides what we have already suggested in this regard, he probably had also in mind the aspect of war where fear flees before courage.[57]

Gandhi's ultimate goal, in a worldwide context, was the promotion and attainment of peace, which he did not feel was ever compromised by his occasional, temporary tolerance of warlike activity. Thus Gandhi stated:

> Whilst I may go with my countrymen a long way in satisfying their need for preparation for war, I should do so in the fullest hope of weaning them from war and of their seeing one day its utter futility.
>
> Let it be remembered that the largest experiment known to history in mass non-violence is being tried by me even as I seem to be lending myself for the purpose of war. . . .
>
> But the war resisters in Europe should strain every nerve to understand and appreciate the phenomenon going on before them in India of the same man trying the bold experiment in nonviolence whilst hobnobbing with those who would prepare for war.[58]

As early as 1921, in fact, Gandhi seems to have come to the conclusion that while believers in nonviolence are pledged not to resort--directly or indirectly--to violence or physical force in defense of anything, they were not entirely precluded from helping individuals, or institutions, that were themselves not based on nonviolence. In support of this stance, he argued:

> If reverse were the case, I would, for instance, be precluded from helping India to attain swaraj,

> because the future Parliament of India under
> swaraj, I know for certain, will be having some
> military and police forces; or to take a domestic
> illustration, I may not help a son to secure jus-
> tice, because forsooth he is not a believer in
> non-violence. . . .
>
> My business is to refrain from doing any
> violence myself, and to induce by persuasion
> and service as many of God's creatures as I can
> to join me in the belief and practice.59 [Empha-
> sis added]

Repeating his familiar argument that, when two parties believing in violence are locked in a conflict, one of them has
justice on its side, Gandhi said that he would be untrue to
his faith if he "refused to assist in a just cause any men or
measures that did not entirely coincide with the principle of
non-violence." In short, he asserted: "My resistence to war
does not carry me to the point of thwarting those who wish
to take part in it. I reason with them. I put before them the
better way and leave them to make the choice."60 Once again
we emphasize that Gandhi did not believe in kindling faith in
people through compulsion.

Another issue in the very last year of his life, namely,
the violation of Kashmir, generated a great deal of contro-
versy over the position Gandhi took in the matter. When
hordes of marauders poured into Kashmir in October 1947
with the apparent incitement and abetment of Pakistan, and
Kashmir frantically appealed to India for immediate military
assistance, the government of India with Gandhi's tacit consent promptly rushed troops for the succor of the ravaged
state.61 Admittedly Gandhi was not very satisfied with the
developments in Kashmir—in fact, he was very distraught
that independence looked like it was going to end in fratri-
cide—but fully appreciated the predicament and posture of
the government of India.

Indications of Pakistan's involvement in the disaster in
Kashmir were already evident well before it actually occurred.
Gandhi's comments thereon, made in his prayer meeting of
September 26, 1947, are rather significant.62 Gandhi said
that he had always been an opponent of all warfare, but if
there were no other way of securing justice from Pakistan, if
Pakistan persistently refused to see its proven error and
continued to minimize it, war would be the only alternative
left to the government of India. While aware of the disastrous
consequences of war, he could never advise anyone to put
up with injustice. And after Kashmir had been attacked, Gandhi said in his prayer meeting of October 29, 1947 that he
could accept the people of Kashmir dying at their post in the
defense of the valley;63 he would not shed a tear if the small

military force provided by India was eradicated like the Spar-
tans while bravely defending Kashmir. His views regarding
warriors, expressed as early as February 1928, that "Fight
Square if You Must" [निदान धर्मयुद्ध करा] were still very
relevant.64

Gandhi, quite naturally, was questioned on the issue:
how could he venture to suggest nonviolence to Hitler, Mussolini, and the Japanese, and not prescribe the same for his
friends in the Congress Government of free India instead of
blessing the troops for Kashmir? His reply was typical: in
neither case had there been any lapse from his creed of non-
violence.65 His ahimsa did not forbid him from giving credit
where credit was due even though the person or institution
to whom credit had to be given--in this case the government
of India--was a believer in violence. He had not personally
advised the use of arms to either of the parties in the Kashmir situation, but he could not, on that account, possibly
withhold his appreciation for the courage and valor of the
defenders. He had no doubt that, if both the relieving troops
and the Kashmiri defenders died heroically, it would perhaps
change the face of India. But if the defense were wholly nonviolent in intention and action, then it would not be necessary to use the label of nonviolence, for the all-round transformation would then be a certainty, even to the extent of
converting the Indian cabinet--if not Pakistan too--to the
viewpoint of those defending Kashmir. Hence his advice on
the Kashmir issue, Gandhi pointed out, was not different
from that he gave Churchill, Hitler, Mussolini, or the Japanese: those who had the courage to be killed to the last soldier in resisting the invader could not put their courage to
a better use.

Fundamentally, Gandhi's attitude to war in general, however, remained unchanged: he was not entirely reconciled
to the war in Kashmir or, for that matter, anywhere
else. Nonviolence, to Gandhi, was the only effective way
open to people and societies; but what could those without
faith in the technique do in the face of aggression? His
philosophy of nonviolence provided no rough and ready escape from their bounden duty to resist evil with all their
might: they had to act.

It is not particularly fair, moreover, to ascribe divinity
or perfection to Gandhi, and make upon him impossible demands of mechanistic logic divorced completely from the context of his situation, values, and objectives. He was, after
all, not infallible, as he himself admitted:

> I ever compromise my own ideals even in indi-
> vidual conduct not because I wish to but be-
> cause the compromise was inevitable. And so
> in social and political matters I have never ex-

acted complete fulfillment of the ideals in which
I have believed.

But there are always times when one has to
say thus far and no further, and, each time the
dividing line has to be determined on merits.

Generally speaking where the sum total of a
movement has been evil, I have held non-co-
operation to be the only remedy and where the
sum total has been for the good of humanity, I
have held co-operation on the basis of compromise to be the most desirable thing.

If I seem to be holding myself aloof from
some of the political movements just now, it is
because I believe their tendency to be not for
the promotion of swaraj but rather its retarding.

It may be that I have erred in my judgment.
If so, it is but human and I have never claimed
to be infallible. . . .[66]

On the whole, Gandhi held steadfastly to his passionate
belief in truth, nonviolence, and the cessation of war. But in
the exceptional cases where nonviolence was not feasible--
due to the parties in the situation being nonbelievers in it
and having been nurtured and trained to fight violently--he
as a realist would not be too eager to advise nonviolence--
the supreme law, "the summit of bravery." To know non-
violence and to practice it in the midst of a world full of
strife, turmoil, and passions was a difficult task as Gandhi
increasingly realized. At the same time, the conviction that
without nonviolence life was not worth living grew equally
deeper in him with every day that passed. He was never a
literalist enough to cling to the letter rather than to the
spirit of an idea, and so long as his own conscience remained
clear he could not be bothered by charges of inconsistency.

Gandhi was not greatly impressed by the stand of the
war resisters, for he felt that mere abstention from war was
not enough; to check war people needed "determination to
suffer to the uttermost in their resistance to war." This was
not evident among the resisters at Saroi, for instance, in
the two peasant wars of Europe that Romain Rolland described.
Similarly, Gandhi protested strongly against the allegation
that he had advised the soldiers to shoot in the air; he asserted:

I do not want a single soldier, after having
taken an oath to serve the army, to mislead the
people by shooting in the air.

I regard myself as a soldier, as a soldier of
peace. I know the value of discipline and truth
and I would consider it unmanly for a soldier

> who has taken an oath to deny himself the con-
> sequences when he defies the orders by shooting
> in the air.[67]

The conclusion becomes apparent, then, that war for
Gandhi is not a phenomenon in isolation. He deals with it as
a seeker of truth. Accordingly, his reaction, his attitude
and views with regard to war have to be studied, not as an
abrupt adoption of the cult of pacifism, but entirely in the
context of the evolution of his all-consuming passion for
truth. His first experience of tyranny revealed that fear was
the very root of all evil. He set about overcoming fear, and
so began his struggle against injustice and tyranny. He set
for himself the goal of truth, which he considered to be far
more valuable than even ahimsa. Truth could never be attained without fearlessness and confidence. Thus, in his view,
courage took precedence over fear, and violence over cowardice; justice became more than mere abstention from violence,
and courage went far beyond mere participation in war. For
the attainment of his goal, he developed the technique of non-
violence to fight an "unjust soulless system." In its dynamic
state, his nonviolence signified--and is meant to even today--
conscious suffering, that is, deliberately inflicting suffering
on oneself rather than doing the same to the opponent. His
concept does not permit meek submission to the will of the
evil-doer, but demands that one's entire soul is pitted against
the will of the tyrant. That is how it becomes possible for a
single frail individual to defy the august might of an unjust
empire, to save his honor, religion, indeed, to save his very
soul.

All his life Gandhi sought truth. In abstract terms, truth
is an absolute value. In concrete, tangible terms its pursuit
came to mean for him the creation of a manifestly just social
order, internally and internationally--something that was not
in the focus, or even serious consideration, of the pacifists
and war resisters. They were preoccupied exclusively with
the avoidance of war. That is, so far as they were concerned,
absence of war constituted peace. Avoidance of war is, obviously, a relative value that "warlessness" is better than
having wars. Consequently, implicit in their position was
natural acceptance of the patently unjust prevailing system
of the world with its power politics, empires, and colonial
exploitation. War was the inevitable function of such an international system. The pacifists and war resisters dissoci-
ated themselves from the symptom but left the disease itself
quite untouched.

Gandhi, in contrast, addressed himself both to the symptom and the underlying disease with equal severity. By demanding India's freedom and due respect of its place in the
world, he challenged and rejected the status quo. Few per-

haps can sustain the argument that empires, with inherent compulsions not only of exploitation of the subject peoples but also those of insatiable rivalry and power politics among the nations owning them, constituted a particularly just world order. In identifying himself totally with India's aspirations for freedom, he unambiguously rejected that order and dedicated himself to the elimination of the oppressive, degrading world of empires with a rare passion. Clearly then, Gandhi drew the value of India's freedom and its natural, inexorable impact on the rest of the world directly from his enveloping environment and realistic social needs.

We must emphasize, therefore, that even though Gandhi was not here expressly concerned in philosophical terms with the larger question of justice, he felt convinced that once the empires made way for a community of cooperating free nations, a just social order would come into being and peace would follow as certainly as dawn follows the night.

His communication to the protagonist of pacifism, Bart de Ligt, is pertinent here: "All activity for stopping war must prove fruitless so long as the causes of war are not understood and radically dealth with."[68] The primary cause of modern wars, Gandhi was sure, was "the inhuman race for exploitation of the so-called weaker races of the earth." Peace has its victories more glorious than those of war, of course, but he was not in favor of "the peace that you find in the graveyard." Gandhi's insistence that peace must be just inheres in it a vision of new world order. Such a peace is a condition in which exploitation and fear find no place and is thus a positive, dynamic, and more enriching and elevating state of affairs than a mere absence of war could ever be. The Western pacifists, obviously, were unable to grasp Gandhi's grand conception of peace and argued with him at cross-purposes, while he proceeded to perfect his position on the issue.

NOTES

1. "I have never been able to reconcile this with his own conduct in other respects, and it is one of the points where I have found myself in painful disagreement." C.F. Andrews, Mahatma Gandhi's Ideas (London: Allen & Unwin, 1949), p. 133.

2. The Collected Works of Mahatma Gandhi, vol. 21 (New Delhi: Government of India, Publications Division), p. 437.

3. Ibid., vol. 21, p. 439.

4. Ibid., vol. 36, pp. 85-86.

5. Ibid., vol. 28, p. 434.

6. Ibid., vol. 36, pp. 108-109.

7. Ibid., pp. 109-110.

8. Ibid., pp. 269ff.

9. Ibid., pp. 85-86; also vol. 37, pp. 269-271; vol. 39, pp. 421-424; vol. 40, pp. 363-365.

10. Ibid., vol. 37, pp. 269-271.

11. Ibid., vol. 39, pp. 423-424.

12. Ibid.

13. Ibid., vol. 40, p. 365.

14. Ibid., vol. 42, p. 437.

15. Ibid., vol. 29, p. 63.

16. Ibid., vol. 37, p. 270.

17. Mahatma Gandhi, An Autobiography, My Experiments with Truth (Ahmedabad: Navajivan, 1962), p. 255.

18. Ibid., p. 257.

19. Collected Works, vol. 36, p. 109.

20. "Everywhere those who oppose violence whether horizontal or vertical are rigorously persecuted. And if Gandhi has been able more or less to gain the confidence of a certain section of the English bourgoisie, it is just because, at certain critical moments, he renounced his pacifism in order to take part in the Colonial and national wars of the British Empire." Barthélemy de Ligt, The Conquest of Violence (London: Routledge & Kegan Paul, 1937), p. 101.

21. International Encyclopaedia of Social Sciences, vols. 11 and 12 (London: Collier-Macmillan, 1972), p. 354.

22. C. Seshachari, Gandhi and the American Scene (Bombay: Nachiketa, 1969), pp. 103-124.

23. R. Niebuhr, Moral Man and Immoral Society (New York: Charles Scribner, 1932), p. 170; also, Richard Fox, Reinhold Niebuhr, A Biography (New York: Pantheon, 1986).

24. Seshachari, Gandhi, pp. 125-136.

25. "To many critics Gandhi's attitude during the World War has appeared to be incompatible with his doctrine of non-violence. Some have interpreted his action as a paradox; while others have looked upon it as a contradiction and a form of opportunism. In his own mystic way, Gandhi made a supreme effort to explain the riddle. Whatever view may be taken. Gandhi's aid to the cause of Allies should make one point clear; Gandhi is not a mere pacifist or conscientious objector." Krishanlal Shridarani, War Without Violence (Bombay: Bhartiya Vidya Bhawan, 1962), p. 115.

26. Ibid., pp. 235-252.

27. Kshitis Ray, ed., "Mahatma Gandhi and Christian Pacifism," in the Gandhi Memorial Peace Number of the Visva Bharati Quarterly (Santiniketan 1959).

28. Paul F. Power, Gandhi on World Affairs (Bombay: Perennial Press, 1961), pp. 48-58.

29. Gene Sharp, "The Meaning of Non-Violence." Journal of Conflict Resolution 3 (March 1959), 41-64.

30. International Encyclopaedia, vols. 11 and 12, pp. 353-356.

31. Collected Works, vol. 40, pp. 363-364.

32. Ibid., p. 363.

33. Ibid., vol. 37, p. 271.

34. Harijan, vol. 8 (New York: Garland, 1973), p. 250.

35. Ibid., vol. 7, p. 327.

36. Ibid., vol. 6, p. 352.

37. Collected Works, vol. 48, p. 402.

38. Ibid., vol. 42, p. 424.

39. Ibid., p. 437.

40. Autobiography, p. 258.

41. Writing in 1921, Gandhi declared: "That is how I would expect every Indian to act by his country under

Swaraj. I should be deeply distressed, if on every conceivable occasion every one of us were to be a law unto oneself and to scrutinize in golden scales every action of our future National Assembly. I would surrender my judgment in most matters to national representatives." Collected Works, vol. 21, pp. 438-439.

42. Ibid., vol. 27, p. 52.

43. Ibid., vol. 37, pp. 270-271.

44. Ibid., vol. 28, p. 226.

45. Harijan, vol. 8, p. 215.

46. Ibid., vol. 7, p. 268.

47. Ibid., vol. 10, p. 477.

48. Collected Works, vol. 13, p. 232; vol. 41, p. 65; vol. 44, p. 114.

49. Harijan, vol. 9, p. 382.

50. Collected Works, vol. 28, p. 434; also, p. 274.

51. Harijan, vol. 7, p. 309.

52. Collected Works, vol. 18, p. 117, vol. 24, p. 140.

53. Harijan, vol. 10, pp. 369, 312. Gandhi also states, "When a policeman comes not to arrest but to molest, he travels beyond his authority. The citizen has then the inalienable right of treating him as a robber and dealing with him as such. He will most decidedly use force in order to defend the honour of womanhood. . . ." Collected Works, vol. 19, p. 117.

54. Collected Works, vol. 18, p. 132.

55. Ibid., vol. 26, p. 489.

56. Ibid., vol. 19, p. 117.

57. War "certainly does one good thing; it drives away fear and brings bravery to surface." Harijan, vol. 8, p. 157. Gandhi also says, "If war had no redeeming features, no courage behind it, it would be a despicable thing and would not need speeches to destroy it." Collected Works, vol. 48, p. 421.

58. Collected Works, vol. 40, p. 365.

59. Ibid., vol. 20, p. 165.

60. Harijan, vol. 9, p. 4.

61. Jawaharlal Nehru, Independence and After (New Delhi: Government of India, Publications Division, 1949), p. 83; Vincent Sheean, Lead Kindly Light (London: Cassell, 1950), p. 201; Pyarelal, Mahatma Gandhi, The Last Phase vol. 2 (Ahmedabad: Navajivan, 1958), p. 696.

62. Harijan, vol. 11, p. 362.

63. Ibid., p. 406.

64. Collected Works, vol. 36, p. 42.

65. Harijan, vol. 11, p. 413.

66. Collected Works, vol. 36, p. 102. See also Autobiography, chap. 38.

67. Collected Works, vol. 18, p. 133; vol. 24, p. 142; vol. 28, p. 434; vol. 36, pp. 25-26; and vol. 48.

68. Ibid., vol. 4, p. 365; vol. 22, p. 104; also Harijan, vol. 8, p. 212; D.G. Tendulkar, Mahatma, vol. 7 (New Delhi: Government of India, Publications Division, 1960), p. 2.

4
EVOLUTION OF GANDHI'S ANTIWAR POSITION

As a young man Gandhi was enthusiastic about the British and his loyalty to the empire absolutely unqualified. His faith in the bona fides of the British government was sincere and firm. His own experience in South Africa revealed to him many ugly truths about the British, of course,[1] but they, at worst, merely bruised and never demolished his implicit, unswerving faith in them. Right up to the beginning of World War I, Gandhi stuck to his illusions about the British and pinned high hopes on the possible place of honor and dignity they would assign to India in their empire. So he aided them unceasingly whenever he found them in distress.

All this changes radically, however, as World War I comes to an end. The wake of the war marks an irrevocable turning point in at least three important aspects: (1) it transformed Gandhi from a loyalist of the empire to an intractable rebel, (2) it enabled Gandhi to play a pivotal role in the mobilization of the Indian people and the concomitant upsurge of nationalism in the country, thereby establishing him as the supreme friend, philosopher, and guide of Indian nationalism, and (3) it set Gandhi on the path to rethinking, revising, and reaffirming his position on war.

The first mortal blow to his faith in the British bona fides came with the passage of the Rowlatt Act, which deprived Indians of their civil rights and sanctioned imprisonment without trial. Gandhi described the act as "a law designated to rob the people of all real freedom."[2] The truth dawned on him that the empire stood for a policy of all-round extortion and exploitation of India. If the act were any indication of the manner in which India was to be rewarded for its sacrifices in the war, Gandhi would have none of such dastardly treachery; there was no hope whatever, under the

circumstances, of genuine freedom for India as a partner in
the British Empire; and "I no longer believe in the empire as
a beneficent power," said the disillusioned, betrayed Gandhi.[3] The history of the Indian subcontinent took a decisive
turn. With courageous candor, Gandhi declared:

> The whole situation is now changed for me. My
> eyes, I fancy, are opened. Experience has made
> me wiser.
> I consider the existing system of government
> to be wholly bad and requiring special national
> effort to end or mend it. It does not possess
> within itself any capacity for self-improvement.
> That I still believe many English administra-
> tors to be honest does not assist me, because I
> consider them to be as blind and deluded as I
> was myself.
> Therefore I can take no pride in calling the
> empire mine or describing myself as a citizen.
> On the contrary, I fully realise that I am a
> pariah, untouchable of the empire. I must, there-
> fore, constantly pray for its radical reconstruc-
> tion or total destruction, even as a Hindu pariah
> would be fully justified in so praying about Hin-
> duism or Hindu society.[4]

The situation called for agitation, and Gandhi gave the
call for satyagraha against the Rowlatt Act. It turned out to
be the first countrywide, mass national movement against the
government. The Jalianwala Bagh (Park) massacre of unarmed peaceful protest assembly on April 13, 1919, and the
brutal humiliation perpetrated upon the civil population of
Amritsar, the town where the tragedy occurred, were the
final blows for Gandhi in his bitter alienation against the
British. It steeled him in his determination to destroy the
alien rule. Indication of that came within a month or so of
the massacre.
Hostilities in the third Anglo-Afghan war broke out in
May 1919. The government claimed that it had to go to war
with Afghanistan in order to save India. Gandhi retorted, "I
would rather see India perish at the hands of Afghans than
purchase freedom from Afghan invasion at the cost of her
honour. . . . To have India defended by an unrepentent Government that keeps the Khilafat and the Punjab [Amritsar]
wounds still bleeding, is to sell India's honour."[5] In total
reversal of his earlier attitude, he thus expressed his absolute refusal to fight any of Britain's wars:

> If another war was declared tomorrow I
> could not with my present views about the

existing Government assist it in any shape or
form.

On the contrary I should exert myself to
the utmost to induce others to withhold their
assistance and to do everything possible and
consistent with <u>ahimsa</u> to bring about its de-
feat.6

It is significant that Gandhi had earlier been training
himself to acquire the capability to resist war, and in the
context of those of the empire, he had felt "that by suffi-
cient service I could attain the power and the confidence to
resist the empire's wars and its warlike preparations. It
would be a good thing for me, who was seeking to enforce
non-violence in my own life to test the extent to which it
was possible among the masses."7 He had even "dreamt of
one day converting [the empire] to methods of peace instead
of war for the sake of even its own existence though in another form."8 But after the Rowlatt Act-Jalianwala phase,
Gandhi finally abandoned these ambitious aspirations as well.

Quite erroneously, Paul F. Power interprets Gandhi's
reaction to war as a mere consequence of the latter's "disillusionment with the British imperialism," whatever that might
mean.9 But such a view is at best nothing more than a half-
truth. For Gandhi had now quite unequivocally expressed his
personal refusal to take part not only in the wars of the empire, but in wars of any and every kind. The columns of
<u>Young India</u> of 1921 are full of Gandhi's categorical declara-
tions in this regard. He asserted, "My position regarding
[the government] is totally different today, and hence I
should not voluntarily participate in its war, and I should
risk imprisonment and even the gallows, if I was forced to
take up arms or otherwise take part in its military opera-
tions."10

This attitude has been interpreted quite incorrectly by
Rheinhold Niebuhr too, who chooses to take a rather restricted view of it, for he says, "Here the important point
is that the violent character of the Government is recognised
and the change of policy is explained in terms of a change
in national allegiance and not in terms of pacifist principles."11 It must be clearly understood, however, that Gandhi had proclaimed his firm resolve of not participating in
any way in any war, <u>even in that of a free India</u>:

I can no longer, in any conceivable circumstances take part in Britain's wars.

And I have already said . . . that if
India attains [what will be to me so-called]
freedom by violent means, she will cease to

be a country of my pride; that time will be a
time for me of civil death.

There can, therefore, never [be] any question of my participation, direct or indirect, in
any war of exploitation by India.[12]

Fundamentally, the fact to be grasped here is that the
aftermath of the war, and not only the impervious but actually vicious attitude of the British government, had irrevocably transformed Gandhi into a nonviolent noncooperator,
who now saw things in a totally different light:

It is no part of the duty of a non-violent noncooperator to assist the Government against war
made upon it by others. A non-violent non-co-
operator may not secretly or openly encourage
or assist any such war. He may not take part
directly or indirectly in it. But it is no part of
his duty to help the Government to end the
war.[13]

Apparently, after deep rethinking on the whole issue of war
and peace, Gandhi had by this time congealed his ideas not
only on war and peace but violence and nonviolence also into
fairly cogent formulations. His biographers, Polak and others
for example, take note of this fact and add that color prejudice, race domination, untouchability, and communal dissen-
sion were some of the other major issues that engaged him
throughout his life. It is after World War I that Gandhi be-
gins to pronounce unequivocally and unambiguously on war.

Defining war as "an ancient method of settling vital affairs of mankind through the arbitrament of the sword," Gandhi wrote:

And so, in 1920, I became a rebel. Since then
the conviction has been growing upon me that
things of fundamental importance to the people
are not secured by reason alone, but have to
be purchased with their suffering.

Suffering is the law of human beings; war
is the law of the jungle. But suffering is infinitely more powerful than the law of the
jungle. . . . Suffering is the badge of the
human race, not the sword.[14]

Earlier, he had had somewhat mixed feelings about the process of war in that occasionally he detected and even admired
some aspects like the discipline, camaraderie, valor, etc., of
the soldiers. After the horrendous death and destruction of
the 1914-18 war, the atmosphere no longer reminds him of

any monastery; he now finds it to be "a terrible whirlwind." Having grown "from truth to truth," Gandhi has come to the conclusion that war is unrighteous, and nothing of value can ever be achieved through it: "The War certainly did no good to the so-called victors."[15] Since he now believes, and repeatedly declares with all the fervor at his command, that war is wrong, his detestation of it grows with the passage of time, especially when he invites us to bear in mind the sheer wastefulness and futility of war. Moreover, he is certain that "Science of war leads one to dictatorship pure and simple. Science of non-violence can alone lead one to pure democracy. . . ."[16] This is what makes him invoke the Allies in World War II to make their choice, and this is what makes him write on July 23, 1939, to Hitler: "Will you listen to the appeal of one who has deliberately shunned the method of war, not without considerable success?"[17]

Gandhi realized only too well now that war was becoming increasingly destructive, barbarous, and inhuman: "No war of which history has any record took so many lives [as World War I]. . . ." Accordingly, his reaction to World II was of far greater horror than ever before, for war per se had become total and bloodier. He confessed, "I was not so disconsolate before, as I am today. But the greater horror would prevent me today from becoming the self-appointed recruiting sergeant that I had become during the last war."[18] Gandhi was, in fact, moved to tears at the mere thought of the possible destruction of his beloved city of London and its great beautiful landmarks like the Houses of Parliament and Westminster Abbey. The progress made by the science of war since the days of the Russo-Japanese War were too devastating in all respects; war was "no longer a matter of personal bravery or endurance." It now aimed at complete annihilation of the enemy country, and nations were busy feverishly making deadlier, far more sophisticated weapons. In fact, "to compass the destruction of men, women and children it might be enough for me to press a button and drop poison on them in a second."[19] Gandhi found the developments in this regard terrifying enough to be condemned with all the emphasis at his command.

He further revised his opinion that war raised the status of a nation. "Past examples have become obsolete," he felt, and now held that "war has become a matter of money and resourcefulness in inventing weapons of destruction." He read What War Means, given to him by Jawaharlal Nehru, with shock and horror, for it contained an account of excesses committed by Japan on China; the discovery that nations indulge in this kind of unbearable, unforgiveable behavior confirmed Gandhi in his abhorrence of war.[20]

Contrary to his earlier view that war brings "courage and bravery to surface" and the impression that "the people's art flowers at the end of a war," Gandhi now says that such an impression

> is half truth. In so far as one people destroy
> another, it is not art but evil which flowers.
> It is to the extent that people willingly suffer
> and sacrifice their lives that art flowers.
> That is why now, at the end of the Great
> War, England and Germany are not progressing;
> on the contrary, poison has spread in them.
> It is true that both have suffered, but the in-
> tention was rather to inflict suffering than to
> suffer. Neither country has cleansed its heart.
> Both, therefore, are again preparing to fight."21

More than the mere evaporation of his innocent romanticism some years earlier, the bitter truths of war now impinge upon Gandhi's sensitivities, obliging him to look at the phenomenon in its stark ramifications. The trauma of the British betrayal, so to speak, contributed in no small way to awakening his sensitivity to the harsher facets of his environment.

THE SWEEP AND MINDLESS WASTE OF WAR

World War I had precipitated unprecedented and excruciating dimensions of destruction and suffering. The more acquainted Gandhi became with the infinite horrors of war, the severity of its impact on individuals and society, the clearer and firmer he got in the formulations of his views and in his revulsion of the entire phenomenon.

More than feeling merely sorry for the hapless masses in the warring nations, he experienced the torment of his soul. The bulk of the people simply did not know what they were fighting for, what was at stake for them on a personal level and collectively as a nation. Only their leaders knew, if that. What wrong had the people of one nation done to those of another? The leaders might possibly be wrong. Why, then, should innocent, unconcerned people, ignorant of the aims and the stakes involved, be dragged into the war to be slaughtered? Gandhi doubted very much if anybody ever spared even half a thought for the vast multitudes the world over. And the plight of the soldiers, in his opinion, was worse still: theirs was not to "reason why;" they did not even know where they were being marched to; they were just cannon fodder. What difference did it make to the dead whether the destruction and devastation is wrought under the label of totalitarianism or in the holy name of liberty and democracy? According to Gandhi even "liberty and democracy becomes unholy when their hands are dyed red with innocent blood."[22] And in the indiscriminate annihilation, he regretted, "the precious inheritance of past ages" was irretrievably lost.

The inhumanity of war percolates down to the individual level, and during its course criminal assaults increase: "Soldiers drunk with the pride of physical strength loot shops and are not ashamed to take liberties with women. The administration is powerless . . . to prevent such happenings."[23] And what unprecedented destruction World War I unleashed! His choice and citation from a journal called The Brotherhood reflect his views in this regard.[24] According to the journal, (1) nations were spending far more on developing weapons and means of vastly increasing destruction than ever before; (2) the capability of these weapons to destroy human beings is far greater than ever before; (3) the intention of nations, accordingly, is to inflict ever greater death and suffering on humanity, rather than "humanizing war" enough to cause minimum pain and suffering; (4) the emergence of ideologies like fascism--which glorify war and attribute to it a purifying and ennobling function necessary for nations--cannot possibly turn people into better human beings and create a more human race; and (5) besides the direct casualties of war, other health problems emerge. This line of thinking was very much after Gandhi's heart, and he went to the extent of finding corroboration in the journal of the British Medical Association, The Lancet, which clearly stated that an increase in venereal disease, influenza, and so forth, was part of modern warfare's grim sequel.

War was also damnable because during its course falsehood flourished at the expense of truth--indeed, all encouragement was for sacrificing the truth. Gandhi found support of this in a book called Falsehood in War Time by Arthur Ponosoby. According to Gandhi, the book "gives a painful record on the part of all the powers that were engaged in the pastime of mutual destruction on the false plea of philanthropy."[25] He thus underscored the fact that during war nothing but truthlessness prevails because wielders of brute force do "not scruple about the means to be used;"[26] their primary objective is somehow to achieve the objective. This, according to Gandhi, was contrary to dharma. And it was all so pervasive: not only do the states at war indulge in excessive and frequent falsehood themselves, but their press also grossly misrepresents facts to the extent that the propaganda becomes a kind of alcoholism of thought for the people of the world. He was substantiated in this by a book he soon chose to publish, Alcoholism of Thought, sent to him by a European friend.

To Gandhi, personally, war was particularly undesirable since it went contrary to and hindered his pursuit of truth: "Truth is the first and the heaviest casualty in war." This is a disastrous thing to happen, in Gandhi's opinion, for once a nation denounces truth it inevitably chooses the path of destruction: "Sacrifice of Truth is the foundation of a nation's destruction."[27]

That force can never resolve disputes had always been Gandhi's firm belief, and the course of events after World War I provided further confirmation: "Force never settles one single problem. Behold the present plight of Europe after such a 'settlement' by force!"[28] Nor does it even solve the original issue at dispute. In fact, violent methods result in mutual annihilation, and by the time the war is over the original issue falls into oblivion, and only the desires and ideals of the victor become paramount. Consequently, argued Gandhi, might becomes right, and the settlement of the original conflict is overlooked in the exploitation of the vanquished. World War I and the subsequent treaties, Gandhi pointed out with forceful reason, have proved it beyond doubt.

Material destruction apart, the most serious damage that war brought about was, according to Gandhi, in the realm of ethics and morals, both individual and national. Thus, besides the unprecedented destruction in World War I,

> Moral loss was greater still. Poisonous forces destructive of the soul [lying and deception] were brought to perfection as much as the forces destructive of the body.
>
> The moral results have been as terrible as the physical. It is yet too early to measure the effect on mankind of the collapse of the sexual morality brought about by the war. Vice has usurped the throne of virtue. The brute in man has for the time being gained supremacy.
>
> The after effects are perhaps more terrible than the actual and immediate effects. There is no stability about the government of any single state of Europe. No class is satisfied with its own condition. Each wants to better it at the expense of the rest. War between the states has now become a war within each state.[29]

Addressing a European audience at Lausanne on December 14, 1931, Gandhi observed:

> The last War, falsely called great, has taught you and humanity many a great lesson.
>
> Human nature during the War did not by any means shine at its best. No fraud, no lie, no deceit was considered to be too much in order to win the War. Foulest charges were flung by a set of partisans belonging to one nation against

another and these were reciprocated with
double vehemence. No cruelty was considered
too great. Nothing was considered base or
mean in order to compass the destruction of
the enemy.

Suddenly as in a flash the friends of yester-
day became the enemy of today. No honour was
safe, nothing was spared, and historians tell
us that there never was so much blood spilled
as during the last War.

This civilisation of the West was weighed in
the balance and found wanting and you have
hardly risen from the deadly effects of that War.

On the contrary, you are slowly and surely
realising the evil effect of war in a more and
more concentrated form. Most of the nations are
on the brink of insolvency--a direct result of
the War.

You are suffering not only from material
bankruptcy but moral, and we are yet too near
the time of the War to be able to measure the
frightfulness bequeathed to us. Nor was evil
confined to Europe. It has broken the bounds
and travelled round to Asia and no one knows
whether he is standing on his feet or head.[30]

What kind of reconstruction, asked Gandhi, could be expected out of such thickening morass? How could normalization
be possible, and what kind of normalization? After colossal
sacrifices of the War, said Gandhi, "we had expected to see
a new Europe as a result of these, expected to see its people more moral and pure, wiser and more God-fearing. But
the evil way which prevailed there, persists even today and
the people who made the sacrifices are in an unhappy
plight."[31] So the war seems to have done no good whatsoever.
Moral bonds have weakened and hatred increased. Indeed,
"the war mentality is no less strong today than it was before
the War and the lure of power and pelf has become stronger,"
asserted Gandhi. This has to be so since the belligerents in
war have been maddened with the passion of destroying each
other at the cost of all ethics. If Germany wanted to dominate Europe, the Allies in the War also desired nothing less
for themselves and sought to crush Germany. "Europe," held
Gandhi, "is no better for Germany's fall. The Allies have
proved themselves to be just as deceitful, cruel, greedy and
selfish as Germany was or would have been."[32]

World War II, Gandhi could see clearly, would be no different. Far from establishing justice of the claims of either
party, both the belligerents would lose in the end, for wars
were not ever won in a real sense. Accordingly, he doubted

very much that the end of the war would manifest any particularly decisive victory. Like the fabled Mahabharata war, World War II would end in mutual exhaustion and the ostensible victors would meet the same fate that awaited the surviving Pandavas. Gandhi's prognosis that Britain and France would emerge with devalued status in international politics from the war was remarkably prophetic:

> Personally I think the end of this giant war will be what happened in the fabled Mahabarata war. . . .
> What is described in the great epic is happening today before our very eyes. The warring nations are destroying themselves with such fury and ferocity that the end will be mutual exhaustion. The victor will share the fate that awaited the surviving Pandavas. The mighty warrior Arjuna was looted in the broad day light by a petty robber.
> And out of this holocaust must arise a new order for which the exploited millions of toilers have so long thirsted. The prayers of peace-lovers cannot go in vain. Satyagraha is itself an unmistakable mute prayer of an agonized soul.[33]

Gandhi wanted nations to learn the lesson of humility from the futile wars and crumbling empires of history: "Napoleon planned much and found himself a prisoner in St. Helena. The mighty Kaiser aimed at the Crown of Europe and is reduced to the status of a private gentleman. . . . Let us contemplate such examples and be humble."[34] Gandhi did not doubt that victory would come to the Allies; "the pity of it is that it will be only so-called . . . [and] it will be assuredly a prelude to a deadlier war."[35] He was, however, utterly disgusted by nuclear blight of the war, and his revulsion and loathing of war in general grew stronger and deeper: "So far as I can see," wrote Gandhi in the columns of the Harijan,

> the atomic bomb has deadened the finest feeling that has sustained mankind for ages. There used to be the so-called laws of war which made it tolerable. Now we know the naked truth. War knows no law except that of might.
> The atom bomb brought an empty victory to the allied arms but it resulted for the time being in destroying the soul of Japan. What has happened to the soul of the destroying nation is yet too early to see.
> Forces of nature act in a mysterious manner.[36]

Gandhi was greatly concerned about the plight of small nations, which suffered the most from war. He was most anxious about the survival of such nations against the threat of their absorption by larger nations in the fever of the war; if not actually absorbed, he apprehended, the small nations were bound to be reduced to the status of mere vassals. He was greatly distressed at the betrayal of Czechoslovakia by Britain and France. His fears deepened since the small nations could scarcely escape the onslaught of fascism and nazism.

> It is clear that the small nations must either come or be ready to come under the protection of the dictators or be a constant menace to the peace of Europe.
> In spite of all the goodwill in the world, England and France cannot save them. Their intervention can only mean bloodshed and destruction such as has never been seen before.
> If I were a Czech, therefore, I would free these two nations from the obligation to defend my country.[37]

In a similar vein, in response to a query as to whether he sould consider himself justified in driving out the British with the help of arms if he had them, Gandhi exclaimed, "Heaven forbid! Contemplate the carnage and misery wrought by the small nations of Europe during the [First World] War."[38]

War was also to be deprecated because it causes acute economic distress, as Gandhi had learned while a resident of South Africa: "I know that during the South African War even all the Republican notes, which 24 hours before the declaration of war were fully worth their face value, became scraps of paper and I understand that such was the case in France and Germany during the [First World] War."[39] The depreciation of money hits the poor the worst and throws the economic life of the nation into a shambles. Trade among countries, because of the inadequate shipment facilities among other things, is also adversely affected, which in turn hurts the entire economy of the nation. Gandhi's cardinal principle of "greatest good of all," rather than that of just a majority, with the masses afflicted by economic distress, also became difficult to realize under the pall of war. All the so-called achievements of war could scarcely undo, or make up, the damage suffered by the economic life and relationships during warfare; the sheer wastefulness of wars was not entirely unknown to Gandhi.

PROVOCATIONS TO WAR

Gandhi held that greed, and the inevitable resort to violence for the satisfaction of that greed, were among the most basic causes of war. He illustrated this by citing examples of the Opium War and the conduct of the British East India Company: "What was the object of the 'Opium War' with China? If China produced her own opium, the Opium War would not have been waged. Students of history know that it was a war of blind self-interest on the part of England."[40] In Hind Swaraj he commented thus on the British East India Company and its designs:

> That corporation was versed alike in commerce
> and war. It was unhampered by questions of
> morality. Its object was to increase its com-
> merce and to make money. . . . To protect the
> latter it employed army. . . .
> Napoleon is said to have described the Eng-
> lish as a nation of shop-keepers. It is a fitting
> description. They hold whatever dominion they
> have for the sake of their commerce. Their army
> and their navy are intended to protect it. . . .[41]

Indeed, Gandhi held the policy of exploitation and commercial advantage in colonial possessions directly responsible for the rise of fascism and nazism and for the outbreak of World War II: "The Nazi power had arisen as a nemesis to punish Britain for her sins of exploitation and enslavement of the Asiatic and African races."[42]

This thread of thought brought Gandhi to his second major cause of war, imperialism. "Is not the prime cause of modern wars the inhuman race for exploitation of the so-called weaker races of the earth?" he pertinently enquired.[43] Talking to an American in 1940, he further queried:

> Why is there war today, if it is not for the
> satisfaction of the desire to share the spoils?
> These large holdings [imperialist posses-
> sions of Western democracies] cannot be sus-
> tained except by violence, veiled if not open.
> Western democracy, as it functions today is
> diluted Nazism or Fascism. At best it is merely
> a cloak to hide the Nazi and Fascist tendencies
> of imperialism. . . .
> It was not through the democratic methods
> that Britain bagged India. What is the meaning
> of South African democracy? Its very constitution
> has been drawn to protect the white man against
> the coloured man, the natural occupant.

> Your own history is perhaps blacker still,
> in spite of what the Northern states did for the
> abolition of slavery. The way you have treated
> the Negro presents a discreditable record.
>
> And it is to save such democracies that the
> war is being fought; there is something very
> hypocritical about it.[44]

Gandhi's whole life became, accordingly, a fight, a totally nonviolent fight, against imperialism, for that was in his thinking the only way to peace.

Although an ardent nationalist himself, about the time of World War II, Gandhi held a brand of nationalism also to be a cause of war. The narrowness, selfishness, and exclusiveness of nationalism made it pernicious and vile in his opinion. Accordingly, he condemned the nationalism of the Italian and German varieties as parochial and vicious, considering fascism and national socialism to be essentially disruptive forces in world affairs. His opinion of Hitler reflects Gandhi's views on nationalism:

> The tyrants of old never went so mad as Hitler
> seems to have gone. And he is doing it with
> religious zeal. For he is propounding a new
> religion of exclusive and militant nationalism in
> the name of which any inhumanity becomes an
> act of humanity to be rewarded here and here-
> after.
>
> The crime of an obviously mad but intrepid
> youth is being visited upon his whole race with
> unbelievable ferocity.[45]

He did not fail to perceive what Mussolini and Hitler stood for, and what they were capable of perpetrating in the name of "nationalism." To Gandhi they represented a form of tyranny that was perhaps unprecedented in history; there was no hope at all of democracy surviving under them. Nor, indeed, did small nations have a chance of independent existence as long as Mussolini and Hitler were in power:

> The philosophy for which the two dictators
> stand calls it cowardice to shrink from carnage.
> They exhaust the resources of poetic art in
> order to glorify organised murder. There is no
> humbug about their word or deed. They are
> ever ready for war. There is nobody in Germany
> or Italy to cross their path. Their word is law.[46]

Apparently, Gandhi had completely revised his earlier appreciation of Mussolini for he had once stated:

It is not that people would necessarily be unhappy under fascism. We may leave aside Hitler,
but under Mussolini Italy is certainly better
off than before. Some of the public utility works
undertaken there are commendable. The standard of living has improved.[47]

Gandhi had also written to Romain Rolland, on December 20,
1931, that behind Mussolini's implacability was a desire to
serve his people; behind his emphatic speeches were a nu-
cleus of sincerity and a passionate love for them; and that
the majority of Italians loved his iron-willed government. But
now Gandhi recognized that the ostensible material advance
under fascism was of no avail since the system completely
denied freedom and was based on force. He felt that these
dictators were trying to show to the rest of the world how
efficient violence could be when not encumbered by hypoc-
ricy or weakness masquerading as humanitarianism. They
were, in fact, showing to the world the hideous, terrible,
and terrifying face of naked violence.

The Soviet nationalism also, Gandhi believed, was similarly not conducive to peace. In fact, Gandhi feared that the
dictatorship there posed an unknown, unimagined menace to
the entire world: "Russia is out of the picture just now.
Russia has a dictator who dreams of peace and thinks he will
wade to it through a sea of blood. No one can say what Russian dictatorship will mean to the world."[48] He had a feeling
that the Soviet dictatorship would create chaos: "Some say
there is ruthlessness in Russia, but that it is exercised for
the lowest and the poorest and is good for that reason. For
me, it has very little good in it. Some day this ruthlessness
will create anarchy, worse than we have ever seen."[49]

Thus nationalism in its narrow confines and selfishness,
in Gandhi's considered opinion, was as dangerous and deplorable as imperialism; indeed, "nationalism otherwise known
as imperialism is the curse," he felt.[50]

And so far as exploitation of the weaker races, people,
according to Gandhi, did not find much difference between
the fascists and nazis on the one hand and the British on
the other. Writing in the columns of the Harijan in 1939, he
maintained:

The democracies therefore that we see at work
in England, America and France are only so-
called, because they are no less based on violence than Nazi Germany, Fascist Italy or even
Soviet Russia. The only difference is that the
violence of the last three is much better organized than that of the three democratic powers.

> Nevertheless we see today a mad race for
> outdoing one another in the matter of arma-
> ments. And if when the clash comes, as it is
> bound to come one day, the democracies win,
> they will do so only because they have the
> backing of their peoples who imagine that they
> have a choice in their own government whereas
> in the other three cases the people might rebel
> against their own dictatorships.[51]

In a written reply to one of Louis Fischer's questions, Gandhi observed in 1942:

> I see no difference between the Fascist or Nazi
> powers and the Allies. All are exploiters, all
> resort to ruthlessness to the extent required to
> compass their end.
> America and Britain are very great nations,
> but their greatness will count as dust before
> the bar of dumb humanity, whether African or
> Asiatic. . . . They have no right to talk of
> human liberty and all else, unless they have
> washed their hands clean of pollution. . . .
> Then, but not till then, will they be fight-
> ing for a new order.[52]

Insofar as exploitation of peoples and areas is, in effect,
economic, Gandhi also duly emphasizes economic factors among
the causes of war. His firsthand experience of the Boer War
and the Zulu uprising perhaps made him put great emphasis
on this factor, and rightly.[53] But he did not accept the Marx-
ian view that economic factor is the sole cause and source of
all evil. It was not correct, he thought, to trace the origin
of all wars to economic causes, nor did he think that wars
were an inevitable consequence of the institution of private
property in capitalist systems. "No, not the economic factor
alone," affirmed Gandhi,[54] "Was not Helen of Troy the cause
of the Trojan War? Were the wars of Rajputs related to the
institution of private property?" He thus pointed out that
sex also is an important factor in motivating war, as the
Illiad, Ramayana, and Mahabharata convincingly illustrate.
Paul F. Powers feels that Gandhi has overlooked "the
motivating factors of prestige and adventure which have been
significant in the history of Western Colonialism."[55] However,
this is not the case. Gandhi, in fact, takes full note of these
factors; only he did not consider them to be particularly relevant in his times, since the predominant aim then seemed to
be money and destruction: "War has become a matter of money
and resourcefulness in inventing weapons of destruction. It
is no longer a matter of personal bravery or endurance." As

a corollary of this, he also held modern machinery responsible for the rivers of blood flowing in war. According to Gandhi, there was hardly any doubt that the "machine age was responsible for the organised murder during the [First World] War. Asphyxiating gas and such other abominations have not advanced us by an inch."56

OPPOSITION TO WAR TAKES SHAPE

Britain and France formally entered World War II on September 3, 1939. Soon after the declaration of war by the British prime minister came the announcement by the viceroy, Lord Linlithgow, committing India to the war without any reference to, or consultation with, the Central Legislature or the representative opinion in India. Such callous arbitrariness on the part of Britain and its viceroy incensed the Indian leadership. The latter strongly felt the humiliation of their country being dragged into a war, not of their making nor of much direct concern to them, without even the courtesy of a formal sounding, let alone consultation and consent. The viceroy's action was particularly galling in view of the fact that (1) self-governing dominions like Canada, Australia, and South Africa had not only been duly consulted but had actually been offered the option of joining the war on Britain's side or staying out of it, and (3) Gandhi and the Congress had seen the war coming and for years had deliberated on the possible stand India should take in the event of war actually breaking out.

The Indian National Congress, from 1929 onward, had adopted resolutions to the effect that in the event of a war it would like to be consulted and would oppose all attempts at involving India in war, particularly when its people had not given their consent. This was the recurring theme of all sessions of the Congress right up to 1939.57 Many years prior to these resolutions, Gandhi had eloquently declared in the columns of Young India his absolute refusal to participate in the wars of the empire even at the expense of being sent to jail or the gallows.58

Indignation apart, the viceroy's declaration posed a very great dilemma for the Indian leadership. Almost all of them were already committed to a full-time struggle for the freedom of India by nonviolent means; national independence was their constant and consuming passion. Now with their country being dragged into war unceremoniously, the question arose concerning what was to be done.

Gandhi would not take the initiative of leading the country. Instead, he merely called upon the nation to decide on the role it should play: "It is in the midst of this catastrophe without parallel that Congressmen and all other responsible

Indians, individually and collectively, have to decide what part India has to play in this terrible drama."59 For the less influential bodies in national politics, the issue was simpler and the course of action rather limited. The Liberal Federation, the All-India Christian Conference, and the Indian Princes, for instance, stood for unconditional, unstinted support to the British for the simple reason that the choice before them was merely between "the frying pan and the fire": between the British with their promises of gradual evolution toward self-government and the Axis powers, which held no such promise. Communal organizations, on the other hand, tended to be motivated by sheer expediency. The Hindu Mahasabha gave its support with the idea of enhancing Hindu representation in the army. The Muslim League assured its full and unconditional cooperation with the expectation of being acknowledged as the only body truly representing the muslims of India, which hopefully would not be ignored in any postwar settlement between the Congress and the British over the political future of the country.

For the Congress, however, the problem of defining the correct stance toward the war was far more complicated and thus difficult. The party had to decide upon two interrelated questions: (1) whether the struggle for India's independence was more important than the war to preserve liberal, democratic values against fascism and nazism or vice versa and (2) whether nonviolence could be used as a technique to meet the challenge of war.

The vast majority of congressmen were skeptical and felt quite unable to apply the technique of nonviolence to the international plane. In fact, the party had often debated in the 1930s the validity of nonviolence in international affairs and had been led to the conclusion that nonviolence as a technique of action was best suited to nationalist struggle within the country; in the realm of international affairs, force had an important role to play. Reflecting this view, the Congress resolution of March 1939 deplored the growth, and aggression, of fascism, while at the same time censured the British policy of appeasement, which had betrayed Britain's treaty obligations and implicitly encouraged the fascist expansion. Indeed, almost on the eve of the war, in August 1939, Congress carried a resolution drafted by Nehru against another resolution--advocating total nonviolence--prepared by Gandhi. The latter was fully aware of the party's attitude in general and its mood of the moment and so at once publicly acknowledged that Nehru's draft represented "more truly than mine, the country's opinion and even the Working Committee's as a whole."60 On September 14, 1939, the party called for a further elucidation of British policy with regard to the war aims and the status of India: if the war were being fought simply to safeguard colonial possessions and imperialist interests, then India would have nothing to do with it; but, on the other hand, if Britain were fighting to save democracies in the world, then it must make India free, and a free India would gladly help Britain against the Fascist and Nazi aggression. In other words, Congress was offering conditional help in the war effort.

Gandhi met the viceroy on September 5, 1939, at Simla and assured him of his personal moral support and sympathies for Britain, while making it clear that he himself could not participate in any way in the war. He was aware, of course, that his sympathy was quite meaningless: "my sympathy had no concrete value in the face of the concrete destruction that is facing those who are directly engaged in the war."[61] How could he with his antiwar stand, and as a believer in non-violence, lend even his moral support? Gandhi's answer was perfectly in character and in keeping with his principles: it was the duty of even one who disapproved of war to distinguish between aggressor and defender. He made it clear that he spoke purely for himself and was acutely aware that with his "irrepressible and out-and-out non-violence" he could not represent the national mind.

In sharp disagreement with Congress, Gandhi regretted that he was "alone in thinking that whatever support was to be given to the British should be given unconditionally."[62] Had Congress not asked for clarification of Britain's war aims, its support would have been unconditional, and thus moral, felt Gandhi. After all, Congress had no soldiers to supply, no money to give; but it had good wishes and, had it offered just those, he thought, it would have made the British cause just. But he would not dream of imposing his views on the party. His position in Congress had become one of a "spiritual director," and he would not change it. Thus, with regard to the war, Gandhi neither determined nor influenced the policy of Congress. Instead, he decided to "plough a lonely furrow, if it is to be my lot that I have no co-sharer in out-and-out belief in non-violence,"[63] though his viewpoint did enjoy some support in the country at large. For instance, a legislator from the Punjab wrote to the Congress president, Dr. Rajendra Prasad, supporting Gandhi's position in this regard.[64] The difference of approach and attitude between the party and Gandhi was, however, to persist throughout the war.

During World War I, when the Congress party was in no mood to cooperate, Gandhi was urging the leadership to help in the war effort.[65] And now, when the party seemed to be keen to help, subject to certain conditions, Gandhi came up with very different ideas. At that time he attended the various war conferences convened by the government to evolve a common strategy with regard to the ongoing war. But now he was not at all inclined to do even that.[66]

The World War II period easily constitutes one of the greatest ordeals in Gandhi's career. His rocklike and all-consuming faith in nonviolence was pitted against the exigencies of practical politics. His passion to see India freed at the earliest time possible could not be consummated as long as the war lasted. In fact, the war had put the prospect of India's freedom in complete jeopardy in many ways. The devastation caused by the war pained him, and he could do nothing about it. This was a trying time, indeed, for Gandhi, and he was very disturbed. According to Maulana Abul Kalam Azad, Gandhi "was so disturbed that on several occasions he even spoke of suicide. He told me that if he was powerless to stop the suffering caused by the war, he could at least refuse to be witness to it by putting an end to his life."67

However, Gandhi philosophically took a personal stand with regard to the war. Unlike the Congress, there was no dilemma whatever for him on the question of association or participation in the war. Nor did he see any question of choice between India's freedom and the saving of the world's democracies. It was perfectly clear to Gandhi that neither of these objectives could be attained by means of war. According to him, in the final analysis the wholesale killing and violence that war unleashed rendered all issues inherent in the two objectives utterly irrelevant. And so, Gandhi resolved not to cooperate in the war, for "the method of non-cooperation and non-violence is not only advisable but one who is convinced of the injustice of war is bound to offer non-violent non-co-operation even if he or she be the only person."68

Gandhi wanted India's independence immediately, of course. His cable of December 3, 1939, to a British Quaker friend, Agatha Harrison, read, "If right men come unofficially they will receive all possible help from Congress. Terrible misunderstanding. Imperialism not dead. No progress possible without unequivocal declaration Independence."69 But as a true satyagrahi, he did not wish to turn his opponents' (be it Britain or Germany) hour of need into his own opportunity: "I am not thinking of India's deliverance. It will come, but what will it be worth if England and France fall, or if they come out victorious over Germany ruined and humbled?"70 Moreover, in contrast to his earlier thinking that war offered a "splendid," "golden opportunity" to Indians for winning a status of equality and respectability in reward for services rendered in the war, Gandhi's revised stance does not permit hankering after such objectives or considerations. He did not now wish to take advantage of Britain's vulnerability and predicament, just as he did not want to exploit Germany's position to further India's goal: "I would not care to erect the freedom of my country on the remains of despoiled Germany."71

The Congress, on the other hand, was prepared to strike a bargain: it would help destroy fascism and nazism on the condition that India be granted its freedom first. The viceroy, on October 17, refused such a stipulation and reiterated the British government's promise of granting India "dominion status" after the war. In its disappointment, Congress turned again to Gandhi for advice. He saw no further point in having the Congress ministries in the provinces continue to function; in the week following the Working Committee's Wardha resolution, of October 23, 1939, the congressmen resigned in protest against British imperviousness. Certain sections of Indian opinion felt that by so doing the Congress, under Gandhi's influence, was doing just what the British government wanted it to do:

> The resignation of ministries . . . will help the Governors and others to work for promoting help. . . .
> You are practically helping the Britishers by resigning. Your moves are nothing else but the manifestation of the commands of the Britishers through the sweet talks and words of your Mahatma sitting on the neck of Congress.
> . . .
> Please be true to the undertaking of the Trust of Indian National Congress and do not be moved by Gandhi and others but only by your true sense of duty to your country.[72]

In any case, once the decision had been made, there was no question of retraction. Gandhi's weapon of civil disobedience now seemed to be the only course for Congress to take. The party at its annual session in mid-March 1940 at Ramgarh resolved to launch the civil disobedience movement throughout the country at a suitable moment under Gandhi's leadership and started building pressure on Gandhi to take the step soon.

This time, however, Gandhi had no intention of a mass satyagraha. Until he was absolutely sure that the country was ripe for it--disciplined and united--and the vicious communal problem would not queer the pitch, Gandhi decided to bide his time. He wrote in Harijan that had the country been demonstrably nonviolent and disciplined he would have promptly launched the civil disobedience, but it was not:

> If the British Government will not suo moto declare India as a free country having the right to determine her own status and constitution, I am of the opinion that we should wait till the heat of the battle in the heart of

the Allied countries subsides and the future
is clearer than it is.[73]

A few months earlier (in November 1939) also he had counseled the impatient congressmen: "I see no immediate prospect of declaring civil disobedience. There can be no civil
disobedience only for the sake of embarrassing Great Britain.
It will come when it becomes clearly inevitable."[74] Gandhi
had then advised them to undertake a constructive program.
They failed, however, to see Gandhi's point. Even the president of Congress, Maulana Abul Kalam Azad, insisted that
his party was not "a pacifist organisation but one for achieving India's freedom. To my mind therefore the issue raised by
Gandhi was irrelevant."[75] Thus the Congress Working Committee meeting at Wardha, on June 21, 1940, decided that
the problem of achieving national freedom had then to be
considered along with the one of its maintenance and national defense against possible external and internal disorder.
If the Nazis were to win the war, India's freedom would be
in jeopardy since in a fascist world freedom to demand would
never be allowed; help, therefore, should be tendered to
save the democracies of the world. Congress, accordingly,
promised total support in this regard provided, of course,
that India was freed. This clearly represented the party's
acceptance of the Nehru-Azad viewpoint. Subsequently, this
attitude crystalized and culminated into what came to be
known as the "Delhi Resolution," according to which the
party formally absolved Gandhi of further responsibility for
Congress. Gandhi, acknowledging a fundamental divergence
of principle with the Congress viewpoint, withdrew from the
party. He could not accept that the strong antifascist fervor
acquiescing to Britain's war aims, held by Nehru and others
in the party, was justification enough to tone down the demand for immediate granting of independence in favor of
conditional support.

Gandhi's reaction to the development was a mixed one.
He reported being "Happy because I have been able to bear
the strain of the break and have been given the strength
to stand alone. Unhappy because my word seemed to lose
the power to carry with me those whom it was my proud
privilege to carry all these many years."[76] But he appreciated the party's stand with regard to the principle of non-
violence. "It is not possible," he acknowledged in the Harijan
dated September 1, 1940, "for a large and popular organisation like the Congress to be wholly non-violent for the simple
reason that all its members cannot have attained a standard
level of non-violence."

What he could not bring himself to accept, however, was
what he considered the Congress's compromise with regard
to war. In spite of all his friendliness toward the British and

his intense aversion to the idea of the triumph of the Axis arms and ideology, Gandhi did not feel it was right to co-operate, even in a nonviolent manner, with the Allies. Still less was he inclined to support the massive campaign of recruitment and war effort already under way throughout the country at an accelerated pace. He was rather disheartened that the Congress seemed willing to assist in a worldwide conflagration. Gandhi knew that the Delhi Resolution had been drafted by C. Rajagopalachari (Rajaji) who had eventually converted Patel to it, and Gandhi regretted that he could not convince Rajaji of the latter's obvious departure from ahimsa:

> He [Rajaji] claims that his very ahimsa has led him to the point which culminated in his resolution. He thinks that I suffer from obsession owing to too much brooding on ahimsa. He almost thinks that my vision is blurred . .. that I advise, as a disinterested but staunch friend, that the British Government should not reject the hand of friendship offered by the Congress.[77]

Gandhi did that, but in his own way and certainly not as a pleader of the Congress case.

In his essay "To Every Briton," Gandhi exhorted the British "to accept the method of non-violence instead of that of war for the adjustment of relations between nations . . . and for cessation of hostilities since war is bad in essence." He called on the viceroy with the plea that Britain should lay down arms and meet Hitler with nonviolence. The Scottish nobleman was literally dumbfounded at what seemed to him to be an outlandish suggestion to the extent of forgetting even elementary courtesies like having Gandhi duly escorted by his aide-de-camp to his car after the meeting!

The British government, however, countered Congress's offer with what came to be known as the "August Offer" made on the eighth of the month in 1940, according to which the government was to set up a Constituent Assembly encompassing all sections of Indian opinion at the end of the war. In the meanwhile the viceroy would enlarge his own council of advisors by including Indians in it.

In its disappointment and depression, the Congress party again turned to Gandhi. The All India Congress Committee (AICC) on September 17, 1940, decided in Bombay to resort to noncooperation, but only under his leadership. As an earnest of its sincerity, the party proclaimed its implicit faith in the relevance of nonviolence not only for winning the freedom of the country but as far as possible for external defense also.

About this time, Gandhi himself thought that now a clash with the authorities was becoming steadily inevitable, as is evident in the cable he sent his Quaker friend, Carl Heath:

> Trying best avoid clash which seems inevitable. Am in communication with Viceroy. Correspondence shows policy enunciated by Amery unchangable. Arrests prominent Congressmen continue. Amery's assurances given you unsupported by patent facts. Congress restraints cannot be used for self destruction.[78]

In another cable to Heath he said, "Most earnest pleading failed. Evidently such is Gods will. Acting with greatest care. Some form of civil resistance inevitable for mere existence."[79] Thus he accepted the mandate of Congress.

THE DYNAMISM OF STUDIED APATHY

Ten days later, September 27, 1940, Gandhi called on the viceroy to inform him of his proposed course of action, which was to invite the Congress leaders to express publicly their opposition to the war, even though he did not wish to embarrass the British government by an antiwar campaign. The viceroy pointed out to Gandhi that that was precisely what Gandhi's contemplated move would amount to, besides, of course, inhibiting India's war effort. In reply, Gandhi said that there was no such possibility. The meeting was inconclusive.

In a masterstroke Gandhi met the viceroy's point without jettisoning his contemplated course of action: instead of launching a mass movement, he inaugurated the individual satyagraha movement. In his press statement of October 15, 1940, Gandhi observed: "I know that India has not one mind. There is a part of India that is war-minded and will learn the art of war through helping the British. . . . Vinoba Bhave [a young trusted devotee of Gandhi] will start the Satyagraha because I know that Vinoba is an out and out war resister. . . ."[80] He made it abundantly clear that he had no intention at that stage of embarrassing the British or bringing harm to the manifest Allied cause in any way.

A couple of days later Vinoba offered the satyagraha by making an antiwar speech at Paunar: it was wrong to help the British war effort with soldiers or money, he said; the only worthy effort was to resist all war with nonviolent resistance. Unlike the earlier noncooperation movements, satyagraha this time was purposely to be restricted to the select few, lest the movement assume a mass character. Its tempo too was kept in a low key lest the war effort be hampered;

the satyagraha remained suspended, for instance, from December 24, 1940, to January 4, 1941, as a gesture of goodwill for Christmas. Satyagraha for the moment meant an assertion of the right to speak against participation not only in the present war but in all war.

According to Gandhi, this kind of civil disobedience was the minimum requirement of nonviolence. He expected the government not to take umbrage over this kind of antiwar propaganda since it was not directed against anybody in particular, and since in the propaganda there was no ill-will against the British nor, for that matter, against the Germans, Italians, or Japanese. In a press interview Gandhi explained that the civil disobedience had been taken up so that

> the world knows that there is at the very least
> a large body of public opinion represented by
> the Congress which is utterly opposed to the
> participation in war, not because it wishes any
> disaster for the British arms, or victory to
> Nazi's or fascist arms, but because it sees no
> deliverance from blood-guiltiness either for the
> victors or for the vanquished, and certainly no
> deliverance for India out of this war.[81]

Gandhi wished all these nations to be at peace with one another, and India, through its nonviolence, to be a messenger of peace to the whole world. He was confident that Harijan would carry enough propaganda against all war as it delivers the message of peace to the world.

With the Japanese attack on Pearl Harbor, on December 7, 1941, and consequent U.S. entry into the war on the side of the Allies, the arena of war spread and the situation became ominous. The Congress Working Committee meeting in Bardoli, on December 23, reviewed the situation since the launching of the individual satyagraha. It reiterated the Bombay Resolution, of September 17, 1940, to the effect that India would help, but only a free India could undertake the defense of the country on a national basis. The committee expressed appreciation of Gandhi's leadership but regreted that Congress could not accept his stance of total nonparticipation in war under all circumstances.

Gandhi stuck to his stance and instead of compromising asked to be relieved of the responsibility that had been thrust upon him by the Bombay Resolution. "In the discussions I discovered," wrote Gandhi to Congress President Abul Kalam Azad,

> that I had committed a grave error in the
> interpretation of the Bombay resolution. I had

interpreted it to mean that the Congress was
to refuse participation in the present, or all
wars, on the ground principally of non-violence.

I found to my astonishment that most members differed from my interpretation and held
that the opposition need not be on the ground
of nonviolence.

On rereading the Bombay resolution I found
that the differing members were right and that
I had read into it a meaning which its letter
could not bear.

The discovery of the error makes it impossible for me now to lead the Congress in the
struggle for resistance to the war effort on
grounds in which non-violence was not indispensable.

I could not, for instance, identify myself
with opposition to war effort on the ground of
ill-will against Great Britain. The resolution
contemplated the material association with Great
Britain in war effort as a price for guaranteed
independence of India.

If such was my view and I believed in the
use of violence for gaining independence, and
yet refused participation in the war effort as
the price of that independence, I would consider myself guilty of unpatriotic conduct.

It is my certain belief that only non-violence
can save India and the world from self-extinc-
tion. Such being the case, I must continue my
mission, whether I am alone or assisted by an
organisation or individuals.

You will, therefore, please relieve me of the
responsibility laid upon me by the Bombay resolution.[82]

Congress leadership found itself divided over the issue.[83]
Party President Azad took the view: "The question before
[Gandhi] was whether we were prepared to take up the position that the Congress would not participate in the present
war on the ground of non-violence alone. We found ourselves
unable to go so far despite our utmost desire to do so." But
the statement issued on January 3, 1942, jointly by the general secretary of the party, J. B. Kripalani, and Dr. Rajendra Prasad, Vallabhbhai Patel, and Prafulla Ghosh seemed to
endorse Gandhi's view:

It would be nothing short of a calamity for the
Congress to abandon non-violence on any account. For by doing so we lose everything in-

cluding what we have achieved for the last 20
years. . . .

Non-violence as the official policy of the
Congress holds even today.

The Working Committee resolution contem-
plates association in the present war in the
remote contingency of the British Government
making an offer acceptable to the Congress.
If that happens we cannot, of course, remain
in the Working Committee.

The signatories to the statement wanted the members of the
AICC to consider the Working Committee's decision rationally
"irrespective of the party loyalty," and added, "We feel that
the Working Committee will welcome the rejection of its resolution if the AICC holds that the contemplated abandonment
of non-violence is against the interests of the country and,
therefore, the Congress principally on that ground should
not participate in the war effort."

Gandhi, however, was once again relieved of the responsibility of leading the Congress at his own insistence. But he
told congressmen that he did not wish the Congress to accept his viewpoint out of compulsion, for then they lose both:
ahimsa because they are incapable of it and swaraj. "Those
of you who think that Gandhi is a good man but it would be
prudent not to go the whole length with him, then you must
accept the resolution. Only those will express their disapproval of it who are sure in their heart of hearts that pru-
dence, political insight, policy and every consideration demands that ahimsa may not be sacrificed for Swaraj."84 So
far as Gandhi himself was concerned, his exit from Congress
leadership did not reduce his commitment to antiwar propaganda. He kept it up

not in the name of the Congress but on behalf
of the resisters of all war on the pure ground
of non-violence, no matter how few the resisters are. It will be kept up for the sake of asserting the right of resisters to carry on propaganda against all war.

They dare not keep still in the midst of the
inhuman slaughter that is going on. They must
not only speak and write against it, but they
must, if need be, sacrifice themselves in the attempt to stop the torrent of blood. . . .

In any case, as a direct consequence of Gandhi's dissoci-
ation with the Congress leadership, the individual satyagraha
movement came to an end. But even as the Congress gave up
using this type of satyagraha, it decided on January 21, 1942,

to take up the implementation of an earlier Gandhian advice, namely, the constructive program in the villages of the country where the war had caused panic and created all kinds of difficulties for the people.[85]

The British debacle in Burma generated tremendous worry and concern in China, and Generalissimo Chiang Kai-Shaik accompanied by his wife made a visit to India, meeting Gandhi in Calcutta on February 18, 1942. The British Government viewed the meeting with suspicion and considerable disapproval. Churchill conveyed to the viceroy that the British government did not quite relish the prospect of Gandhi and Nehru on the one side, the viceroy of India on the other, and Chiang Kai-Shaik arbitrating between them.[86] But Churchill need not have feared such an occurrence. Gandhi and Chiang Kai-Shaik did not discuss the idea of noncooperation, since the latter did not find it of any relevance to countries other than India! The generalissimo just could not reach Gandhi's level. In his letter to Patel, Gandhi gave his impression of the meeting:

> He [Chiang Kai-Shaik] came and went without creating any impression, but fun was had by all.
>
> I would not say that I learnt anything, and there was nothing that we could teach him.
>
> All that he had to say was this: Be as it may, help the British. They are better than others and they will now become still better.[87]

The spread of the war to the Pacific and the Japanese conquests in Southeast Asia placed India in a very strategic position. President F. D. Roosevelt thought that better cooperation could be had from India if it were assured of its freedom. So did the Labour leaders in the British war government. Accordingly, Sir Stafford Cripps, a Labour member of the British war cabinet, was sent to India on March 23, 1942, with a prepared, unalterable plan to settle the Indian problem. The Cripps Plan, as it came to be known, was not negotiable: it had to be accepted, or rejected, in its entirety. Cripps put the proposals of the British government before the leaders of all the major Indian parties. Discussions with Congress were conducted mainly through Nehru--a personal friend of Cripps--and Maulana Abul Kalam Azad, the president of the party.

With Gandhi, Cripps discussed the proposals only once. Gandhi had made it clear that he represented only himself, not the Congress or anybody else. Gandhi's reaction to the proposals, as he later told Louis Fischer, was: "I said to Cripps why did you come if this is what you have to offer? If this is your entire proposal to India, I would advise you to take the next plane home."[88] Cripps did not take the next plane home, of course, but stayed on for his rounds of "negotiations," though there was precious little to negotiate; Indians could take it or leave it as the plan was. Instead Gandhi left for Sevagram soon after his meeting with Cripps, with firm advice to Congress to make its own decision. He left Congress free since he knew that many a leader in it wanted to participate in the conduct of war; his nonviolence would not admit of compelling, or embarrassing, even a single such person.

Gandhi's disapproval of the Cripps Plan arose from his antiwar stand. When he did not wish to participate in the war, how could he then accept the provision in the plan according to which the British government was to retain control over the defense of the country with Indians joining in? Gandhi could not, even hypothetically, accept the proposition that the real control over India's defense be retained by Britain instead of being duly given over to responsible Indian ministers. Congress certainly did not accept it, and Cripps had to return home empty-handed.

Mention must be made of the earnest efforts that Gandhi's Quaker friends in England, especially Horace Alexander and Agatha Harrison, were making to bring about some understanding between the British government and the Indians. When the Cripps mission was announced (late February-early March 1942), the Quakers wrote to Gandhi, reminding him of "Andrews legacy," meaning that the best Englishmen and best Indians should work out some understanding as a tribute to the memory of Deenabandhu C. F. Andrews. Following the failure of the Cripps mission, Gandhi wrote to Horace Alexander:

> Sir Stafford Cripps has come and gone. How nice it would have been if he did not come with that dismal mission.
>
> I talked to him frankly as a friend, if for nothing else, for Andrews' sake. I told him that I was speaking to him with Andrews spirit as my witness.
>
> I made suggestions, but all to no avail. As usual, they were not practical.
>
> I had not wanted to go. I had nothing to say being "anti-all wars". I went because he was anxious to see me. All this I mention in order to give the background.
>
> I was not present throughout the negotiations with the Working Committee. I came away. You know the result. It was inevitable. The whole thing has left a bad taste in the mouth.
> . . .

>My firm opinion is that the British should
>leave India now in an orderly manner and not
>run the risk that they did in Singapore, Malaya
>and Burma. The act would mean courage of a
>high order, confessions of human limitations and
>right doing by India.89

Concerned intensely as he was with world peace, Gandhi
genuinely wanted India to make a positive contribution toward
it. He found it humiliating, however, "for India--a big nation
--to be not able to help, although it is conscious that it can
render inestimable help in a variety of ways such that would
ensure victory for Allies." To be able to do that, India must
absolutely be free, he insisted: "Unless we are free, we can
do nothing."90 He consistently stuck to this position all
through the months following the Cripps mission, when with
the fall of Rangoon in March 1942 Japan was virtually knocking at the gates of India. He fully appreciated the danger of
a possible Japanese invasion of India, of course, but asserted nonetheless that India would do everything to prevent
Japanese domination--that is, a free India.

In the evolution, and buttressing, of his antiwar posture,
the Cripps mission certainly had a significant place. It is
after the failure of the mission that Gandhi's conviction deep-
ens on two interrelated points: only a free India can make a
positive contribution toward ending the war and ushering in
world peace, and the British should withdraw from India immediately. It was a Gandhi very different from his "individual
satyagraha" days who in his demand for immediate British
withdrawal from India argued in April 1942: "If the British
left India to her fate . . . non-violent India would not lose
anything. Probably, the Japanese would leave India alone
whatever the consequences. . . . Her real safety and of
Britain's too lies in orderly and timely British withdrawal."91
In the same vein, he told Louis Fischer in June 1942 that
there must be civilian control of the military in wartime; had
India been given charge of its defense, it could have done
things that would have been more conducive to success in
the war. In an obvious reference to the provision of "dominion status" for India at the conclusion of the war contained
in the Cripps Plan, Gandhi said that he was not interested
in future promises--the widely reported version of Gandhi's
attitude was that Cripps's proposals amounted to "a post-
dated cheque on a crashing Bank." He wanted immediate independence of India, for it would help Britain also. His appeal "To Every Briton" containing the demand for immediate
withdrawal from India had already appeared a month earlier
(in May).

These were not easy times for Gandhi. Congress was not
in agreement with him on the issue of nonparticipation in the

war. Congress leaders like Rajaji, in face of strong disapproval of his colleagues like the party president, Azad, general secretary Kripalani, and Nehru, were still trying to reach some kind of understanding with the British by first recognizing the Muslim League. Gandhi appreciated Rajaji's move but suggested that he should resign from Congress first before pursuing his efforts.[92] Azad and Nehru had reacted sharply against Gandhi's draft of the resolution-- sent through Mirabehn--demanding immediate British withdrawal from India, for they felt that such a step will go in Japan's favor. According to J. B. Kripalani, Nehru and Azad's opposition to the draft was so strong that they even threatened to resign. The result was that in the modified Congress resolutions the operative part of Gandhi's draft-- namely, the one asking for immediate British withdrawal-- was omitted, even though Kripalani himself, Rajendra Prasad, and Patel were in agreement with Gandhi on the issue.[93] Nehru felt--and Azad agreed with him--that in the critical international situation Gandhi was taking a rather nationalistic stand that completely ignored the crucial international considerations.[94] Gandhi wanted both Nehru and Azad to resign from the Working Committee on this issue.[95] However, they did not, and the debate in the party continued. Eventually, at Gandhi's behest, the AICC decided in August 1942 to launch a massive civil disobedience movement.

Gandhi was absolutely uncompromising at this stage; so much so that he withdrew his moral sympathies for the British to express them eloquently for China and the Soviet Union. His sympathy for the Soviets seemed to be new, for earlier he had been quite skeptical, to say the least, about their designs and role in international affairs. On the Russo-German Pact of August 23, 1939, his comment had been, "Though the part Russia is playing is painful, let us hope that the unnatural combination will result in a happy though unintended fusion whose shape no one can foretell."[96] After Germany's invasion of the Soviet Union in June 1941, Gandhi wrote, "And now that the war has broken out between Russia and Germany, we are unable to say what the ultimate result will be," and left it at that.[97] The Indian communists tried to persuade Gandhi that Hitler's attack on the Soviet Union had altered the character of the war and made it into a "peoples' war." He was not convinced, however, as shown by his correspondence with the general secretary of the Communist Party of India, P. C. Joshi:

> I understand that although the chief actors
> among the Allied powers are by no means in-
> clined towards real democracy, you think that
> by the time the War ends their designs will be
> confounded and that the people all the world

over will suddenly find self-expression and
overthrow the present leaders.

In the peoples, I am entitled to include us,
other Asiatics and Negroes, for that matter
perhaps also the proletariats of Japan and Germany.

If such is your belief, I must confess that I
do not share it, but I keep myself open to conviction.

Meanwhile, I suggest that the title "peoples'
war" is highly misleading. It enables the government of India to claim that at least one popular
party considers this as peoples' war.

I suggest too that Russia's limited alliance
with the Allies cannot now by any stretch of
imagination convert what was before an imperialist war against the Nazi combine into a peoples'
war.[98]

In May 1942, however, he came out with the statement that
his sympathies were "undoubtedly in favour of China and
Russia." Is it because he had always to side with the under-
dog, the aggrieved, and the wrongly injured? About Britain
he now said the following:

I used to say that my moral support was entirely with Great Britain.

I am sorry to have to confess that today my
mind refuses to give that moral support. The
British behaviour towards India has filled me
with great pain. I was not quite prepared for
Mr. Amery's performance or Sir Stafford Cripps
mission. These have, in my estimation, put
Great Britain morally in the wrong.

And, therefore, though I do not wish any
humiliation to Britain and, therefore, no defeat,
my mind refuses to give her any moral support.[99]

So Gandhi decided that Britain must leave India immediately for a variety of reasons--including that the British
presence in India constituted the greatest obstacle to communal harmony in the country; indeed, it incited and exacerbated the tension and discord between Hindus and Muslims.
Once the idea gripped him, Gandhi bent all his energies to
building the "Quit India" movement. The government did not
fail to see how determined Gandhi was: "Present indications
are that he will throw off all pretentions of non-embarrass-
ment [and] declare himself openly anti-British."[100]

THE QUIT INDIA MOVEMENT

In his own mind Gandhi was absolutely clear that the "Quit India" movement was perfectly in line with the proclaimed British war aims, and interests, rather than being inimical to them--for Britain was fighting for democracy. Failure on Britain's part to appreciate this simple fact led him to conclude that both British and U.S. claims of saving democracy were totally unsubstantiated and unwarranted, for here lie a terrible "tragedy of holding a whole nation in bondage." He implored the United States to see that justice was done and India granted unconditional independence forthwith. To say "but this is not the time" is wrong; indeed, "this is the psychological moment for that recognition. For then, and then only, can there be irresistible opposition to Japanese aggression." Actually, the United States should insist upon immediate Indian independence as a precondition for financing the British war effort: "he who pays the piper has the right to call the tune." Washington should "look upon the immediate recognition of India's independence as a war measure of first class magnitude." As to why he had chosen that particular moment for launching the Quit India movement, his reply was:

> India is not playing any effective part in the War. Some of us felt ashamed that it is so. . . .
> If we were free from foreign yoke, we should play a worthy, nay, a decisive part in the World War. . . .
> To distrust this simple, natural and honest declamation was to court disaster.[101]

He was asked: "What is the difference between Nero and yourself? Nero was fiddling when Rome was burning. Will you be also fiddling in Sevagram after you have ignited the fire which you will not be able to quench?" Gandhi replied caustically:

> The difference will be known if match, if I have ever to light it, does not prove a "damp squib". Instead of fiddling in Sevagram you may expect to find me perishing in the flames of my own starting if I cannot regulate or restrain them.
> But I have a grouse against you. Why should you shove all the blame on to me for all that may happen by reason of my taking action for the discharge of an overdue debt and that, too, just when the discharge has become the necessary condition of my life? . . .

It is misuse of language to liken to the action
of Nero that of a man who, in order to escape
living death, lights his own funeral pyre to
end the agony.

Gandhi launched the Quit India movement notwithstanding
the hostility, and misrepresentation, by the press of England
and other nations that he was trying to take advantage of
Allied reverses in the war. He was arrested on August 9,
1942. On the eve of his arrest he had given the nation his
message of "do or die." The vast multitudes that promptly
responded, through means nonviolent or violent, provoked
the government to unleash severe repression. Total censor-
ship of press and media was instituted; Harijan of August 16,
1942, was forfeited, and the journal did not appear until the
end of the war. Gandhi was branded pro-Japanese and a
"fifth columnist."

Gandhi was hugely amused by such ridiculous libel. In
his essay "To American Friends" he wrote that "a posture of
pro-Japanese or 'Fifth-columnist' was contrary to [my] creed.
. . . Having imbibed the message of Unto This Last I could
not be guilty of approving of Fascism and Nazism, whose cult
is suppression of the individual and his liberty." Even General Smuts, Gandhi's erstwhile opponent in South Africa, was
moved to retort indignantly: "It is sheer nonsense to talk of
Gandhi as a fifth-columnist. . . . He is the last person to
be placed in that category."[102]

Gandhi used his own technique to emphasize the sincerity
of his purpose: he decided to go on a fast and conveyed to
the government, "the law of Satyagraha as I know it prescribes a remedy in such moments of trial. . . . it is, cru-
cifying the flesh by fasting."[103] In the extensive correspondence that passed between Gandhi and the government subsequently, the viceroy chose to consider his resort to fasting
as "political blackmail." Anyway, his fast in imprisonment
commenced on February 9, 1943, and lasted for 21 days.

The British government, naturally, misunderstood the
significance of the fast and, except for the Labour members
in it, would perhaps not have been particularly sorry had
Gandhi died of his fast.[104] Gandhi survived the ordeal, but
at what cost! He lost his secretary and confidant, Mahadev
Desai, and the "teacher" of his nonviolence, Kasturba, his
wife. If Churchill remained suspicious still of the bona fides
of his fast, the viceroy, Lord Wavell, also doubted Gandhi's
sincereity, particularly since he felt antagonized even before
he took over as the viceroy in succession to Linlithgow. According to Penderel Moon:

Wavell disliked Gandhi and wholly distrusted
him. . . . He was strongly influenced in his

recollection that in 1942 when he as Commander-
in-Chief was trying to secure India against
Japanese invasion Gandhi had let loose the "Quit
India" rebellion which for some weeks paralysed
communications with the eastern front.
Wavell conceded later that Gandhi had not deliberately intended this as a stab in the back,
but this is what in effect it was.[105]

The charge of Gandhi being a "fifth columnist" and pro-
Japanese stemmed possibly from the activities of Subhash
Chandra Bose and his Indian National Army (INA). But nothing could be more wrong than reading any kind of liaison
between the two. There had been a fundamental difference
of approach to the question of India's independence between
Gandhi and Subhash. The former had never been impressed
by what some considered to be Subhash's "fascistic" tendencies and had had him expelled from Congress soon after the
Tripura Congress in early 1939. Perhaps it was irrepressible
impatience on the part of Subhash, with regard to the country's freedom, that was responsible for him being labeled
"fascistic;" he was all for using every possible opportunity
that offered itself for promoting the cause of Indian independence, rather than being inhibited by considerations like
British embarrassment or individual moral tenets. According
to Subhash, April 23, 1939, was to be observed as antiwar
day when public opinion in the country was to be mobilized
for nonparticipation in an imperialist war.[106] After the war
broke out, he--as the leader of the Forward Block--expressed
his skepticism openly about Gandhi's method and stand on the
issue. In January 1940 he issued the following statement:
"The political significance that is being given to spinning
[charkha] now and the manner in which it has been quietly
converted by the Congress High Command into a method of
political struggle need unequivocal condemnation."[107] About
the same time, January 15, Gandhi wrote to Deenabandhu
Andrews: "Subhas is behaving like a child. . . . His politics
show sharp differences. . . . The matter is too complicated
for Gurudev [Rabindra Nath Tagore] to handle. Let him trust
that no one in the [Working] Committee has anything personal
against Subhas. For me, he is my son."[108] Later in the year
at Subhash's arrest Gandhi expressed the view:

Subhasbabu has laid out his plan of battle. He
thinks that his way is the best. . . . He told
me in the friendliest manner that he would do
what the Working Committee [Congress] had
failed to do.
He was impatient of delay. I told him that,
if at the end of his plan there was Swaraj dur-

ing my lifetime, mine would be the first telegram of congratulations he would receive. If,
while he was conducting his [campaign] I
became a convert, I should whole-heartedly
acclaim him as my leader and enlist under his
banner.
But I warned him that his way was wrong.[109]

Subhash seemed to pin considerable faith on the propa-
gandist professions of sympathy, and possible help, by the
Axis powers for Indian aspirations. Gandhi had no faith in
them whatever:

I have never attached the slightest importance
or weight to the friendly professions of the
Axis powers.
If they came to India, they will come not as
deliverers but as sharers in the spoils.
There can, therefore, be no question of my
approval of Subhas Babu's policy.

Gandhi had great admiration for Subhash's patriotism, of
course, but thought him to be misguided. So, on being told
that people felt happy when Subhash said on the Axis radio
that there were no differences between him and Gandhi, the
latter's indignant rejoinder was: "I do not feel flattered when
Subhas Babu says I am right. I am not right in the sense he
means. For he is attributing pro-Japanese feelings to me."[110]
Without doubt, by this time Gandhi's sneaking but not particularly articulated admiration for Japan and its system, implicit in his writings about the Russo-Japanese War (1904-
05), had quite evaporated. Now he not only had serious reservations about, but positively disapproved of, Subhash
Bose's idea of freeing India with the help of Germany and
Japan--not that he ever had even entertained the thought of
outside assistance in freeing India. In fact, all his life he
had stressed that India must win its freedom entirely by its
own efforts, by the moral, unique technique of ahimsa and
satyagraha; hence his constant urgings for moral regeneration of the people of India. As a corollary of his thorough
disapproval of Subhash Bose's approach, Gandhi did not view
with particular favor the activities of the "Indian National
Army" raised by Subhash in Singapore and Burma in 1943-44.
On being asked about his reaction to the eventuality that
Subhash Bose made a treaty with Germany and Japan, under
which India would be declared independent with the Axis
powers entering the country to drive out the British, Gandhi
declared:

> Surely there is as much difference between the
> South Pole and the North as there is between
> the imagined conditions. My demand deals with
> the possessor; Subhasbabu will bring German
> troops to oust the possessor. Germany is under
> no obligation to deliver India from bondage.
> Therefore Subhasbabu's performance can only
> fling India from the frying pan into the fire.

Again,

> If I were to discover that by some strange mis-
> calculation I had not realised the fact that I was
> helping the entry of the Japanese in this coun-
> try, I should not hesitate to retrace my steps.
>   As regards the Japanese, I am certain that
> we should lay down our lives in order to resist
> them as we would to resist the British.[111]

In his own way, Gandhi was preparing the nation to offer
nonviolent resistance to the invading Japanese, whereby they
were to be denied all possible assistance, even water. Gandhi
was not bothered about the criticism of his Quit India movement by the British and the U.S. press, which flung all kind
of abuse and fabrication at him constantly, rather than trying
to understand his viewpoint. But the Indian press "listening
to the British Censor" did worry him, and "to resist that
awful atmosphere" he gird up his loins.

The ignorance--indeed, sheer prejudice--of the U.S.
press about the ideas and style of Gandhi with regard to the
war was nothing short of amazing. U.S. journalists were just
unable to appreciate his way of facing an invasion nonviolently. They could appreciate a Churchill making a temporary
surrender of England to Nazis as a tactical measure and setting up a government-in-exile, as General Charles de Gaulle
had done by heading a French government based in London,
of course. In fact, Churchill had actually braced President
Roosevelt for such an eventuality.[112] But the Gandhian approach to a possible Japanese invasion of India seemed to be
beyond U.S. comprehension.

Gandhi's approach, and style, was to inspire and pre-
pare his unarmed countrymen to offer nonviolent resistance
to the last person rather than submit to the invading Japanese. His replies to queries from Mirabehn make his approach
in this regard abundantly clear.[113] He wanted his countrymen to keep calm, even in the face of the British reverses
occurring, rather than panic. He advised women to shift to
villages where there was relatively greater security against
the unwarranted and brutish liberties of conquering soldiers.
Additionally, Gandhi urged the press to conduct a sustained

campaign of educating soldiers, "white or brown," and their officers against misbehavior toward women. Lastly, he exhorted his countrymen to make a genuine contribution to the war effort by saving on food and clothing, which were in short supply.114

At the same time, Gandhi left no doubt in the minds of the Japanese that they would not get any welcome from India. He had no illusions about the Japanese or their interest in India having its due, nor did he really care for their dubious help in freeing India from the British yoke: "It is a folly to suppose that the aggressors can be benefactors. The Japanese may free India from the British yoke, but only to put in their own instead."115

Moreover, now his antiwar sentiment was so strong that even for the attainment of his most cherished goal--India's independence--he would not dream of seeking the help of one of the parties to the war.

With regard to Nehru's stated advocacy of guerilla warfare against the Japanese, Gandhi was not impressed and did not take it very seriously:

> I am sorry that he has developed a fancy for the guerilla warfare. But I have no doubt that it will be a nine days wonder. It will take no effect. It is foreign to the Indian soil.
>
> Twenty-two years' incessant preaching and practice of non-violence, however imperfect it has been, could not be suddenly obliterated by the mere wish of Jawaharlal and Rajaji, powerful though their influence is.
>
> I am, therefore, not perturbed by the apostacy either of Jawaharlal or Rajaji. They will return to non-violence with renewed zest, strengthened by the failure of their effort.

Moreover, Gandhi maintained:

> Neither goes to violence for his belief in it. They do so because they think probably that India must have a course of violence before coming to non-violence. No one can say beyond doubt how events will shape themselves. It may be that their instinct is correct, and mine backed though it is by experience, is not.
>
> Guerilla warfare can take us nowhere. If it is practised on any large scale, it must lead to disastrous consequences. Non-violent non-co-operation is the most effective substitute for every kind of violent warfare.116

During 1942-44 Gandhi was quite articulate with regard to the twin issues of the Japanese invasion and the permissibility of the Allied troops to operate from bases on the Indian soil.[117] His observations were colored entirely with his irrepressible keenness to see India free at the earliest moment possible during the war, rather than at its conclusion. The struggle for independence perforce was influenced by the course of the war, especially when it started knocking at the gates of India. He was satisfied that "the British presence is the incentive for Japanese attack. If the British wisely decided to withdraw and leave India to manage her own affairs in the best way she could, the Japanese would be bound to consider their plans."[118]

There was logic in Gandhi's insistence on immediate British withdrawal that would leave India independent. British presence in the country, he could see, was quite obviously acting as a bait for the Japanese to attack; once Britain left the shores of India, the country would cease to attract Japanese belligerency.

More important than that, however, was the moral aspect of the move. By withdrawing from India, the British would automatically put the entire Allied cause on a completely moral footing, which may well lead to a most honorable peace between the warring nations. By holding India in bondage, Britain was behaving no differently than Nazi Germany, Fascist Italy, and a Japan riding the rhetoric of "freeing" Asia of Western occupation and exploitation. These powers were now building empires, for Britain and its allies had them.

Further, an India in bondage could never properly manifest active hostility to any Japanese advance. A free India would have no interest or motive whatsoever to welcome the Japanese, nor even to declare a war against. Instead, "she simply becomes the ally of the Allied powers, simply out of gratefulness for the payment of a debt however overdue," for Gandhi believed, "human nature thanks the debtor when he discharges the debt."

So Gandhi's logic was simple: with British withdrawal from India the bait for Japanese attack would have been removed; a grateful free India would willingly cooperate with the Allies whose cause would automatically become moral and thus effective for ensuring a quicker restoration of peace in the world. In short, the Japanese could be kept effectively out of India only if the British immediately withdrew. And he assured a U.S. journalist: "Remember, I am more interested than the British in keeping the Japanese out. For Britain's defeat in Indian waters may mean only the loss of India, but if Japan wins India loses everything [emphasis added]."[119]

Simultaneously, Gandhi was heavily burdened with the thought that he had the responsibility of preventing the Japanese from reaching China and the Soviet Union, via India.

In the discharge of this responsibility, he was even willing
to cooperate with the British. In an interview he gave to
Louis Fischer in the summer of 1942, he outlined the policy
and course of action of the National Government of Free India:

> The first act of the National Government would
> be to enter into a treaty with the United Na-
> tions for the defensive operations against the
> aggressive powers, it being common cause that
> India will have nothing to do with any of the
> fascist powers and India would be morally bound
> to help the United Nations.[120]

Only a couple of months earlier, in April 1942, he had expressed his disapproval of the advent of U.S. soldiers in
India. Also in June he argued:

> We know what American aid means. It amounts
> in the end to American influence, if not Ameri-
> can rule added to British.
>   It is tremendous price to pay for the possible
> success of the allied arms. I see no Indian free-
> dom peeping through all this preparation for the
> so-called defence of India. . . .
>   Holding the views I do, it is clear why I
> look upon the introduction of foreign soldiers
> as a positive danger thoroughly to be deplored
> and distrusted. . . .[121]

So great was his concern for the security of China, however,
that he now agreed to allow the Allied troops to stay in India, for their "abrupt withdrawal . . . might result in Japan's occupation of India and China's sure fall."
    This shift in Gandhi's position has to be seen in the
total perspective of the situation then obtaining not only in
India but the world. Gandhi was in no position to keep the
Japanese at bay by his nonviolent action for the simple reason that he knew only too well that the whole of India had
not yet acquired proficiency in nonviolence; how could he
then guarantee success of his technique in the context of the
world situation in 1942? Secondly, in return for his consent
to allow the Allied troops in India, Gandhi expected to invoke the moral obligation of the British to declare at once
the freedom of India. Finally, the sheer helplessness of a
country held in bondage also could not have escaped his
perception of the situation: "till we come to our own," he
had written as early as 1929, "we shall have to be victims
of the war that may come upon the world."[122]

Over and above is his overriding concern to bring about the end of the war in Asia, and for that Gandhi seems willing to stretch a point. In fact, on his release from imprisonment in 1944 he does not even seek control of the Indian defense by the national government, and assures that he will not interfere--let alone prevent--in the national government opting to participate in the War. All this is subject, of course, to India being declared free first.[123]

Gandhi spelled out distinctly the implications of his willingness to allow the Allied troops in India: (1) India would be free of all past and present financial obligation to Britain; (2) the annual drain to Britain, with regard to economics and services, would stop automatically; (3) all taxation lapses, except the one that the new government would impose or retain; and (4) the dead weight of an omnipotent authority over the most powerful in the land would be lifted at once. In short, India would begin a new chapter in its national life, and Gandhi hoped to affect the fortunes of the war with nonviolence as his primary sanction. This nonviolence, rather than being nonparticipative or noncooperative, would express itself by the ambassadors of a free India going to the Axis powers not to beg for peace but to show them the utter futility of war for achieving honorable ends.[124]

Gandhi had no doubt that his views, and of course the specific proposals, had a wider relevance--not only to India but also to the whole world, not only to those engaged in war but also to those who were not. He could not bear with equanimity, or indifference, the butchery going on in the world; the nonviolent war resister in him revolted at the wasteful, senseless, insane, debasing spectable. He earnestly hoped and trusted, therefore, that the acceptance of his proposals--and their full fruition--would change the character of the war:

> What is today a war of brute strength, would be turned into a war for the liberation of the exploited peoples of the world.
> Then it would be a war between predominantly moral strength, plus the minimum of brute strength, pitched against pure brute strength which is being used for the exploitation of China and the weaker states of Europe.[125]

Gandhi was convinced that the acceptance and implementation of his proposals would at once transform the Allied cause into something totally different, something purer, compared to the "cause"--if ruthless annihilation could be so described--of the Axis powers.

His letter of July 27, 1944, to the viceroy, Lord Wavell, truly reflects the shift in Gandhi's position vis-à-vis his as-

sociation with the war effort, and the role and responsibilities of the proposed national government. The letter assured
the viceroy that another civil disobedience movement would
not be launched "in view of the changed conditions," provided of course that "a declaration of immediate Indian independence is made." Furthermore, Gandhi asserted:

> I am prepared to advise the [Congress] Work-
> ing Committee that, in view of the changed
> conditions, mass civil disobedience envisaged
> by the resolution of August 1942, cannot be
> offered and that full co-operation in war effort
> should be given by the Congress if a declara-
> tion of immediate independence is made and a
> national government responsible to the Central
> Assembly be formed subject to the proviso that
> during the pendency of the war, the military
> operations should continue as at present but
> without involving any financial burden on In-
> dia. . . .126

The Tory press in England, misinterpreting this significant gesture by Gandhi, described it as insincere, hypocriti-
cal, and incompatible with his basic creed of nonviolence,
for the furtherance of the war effort, in their view, was
quite inconsistent with nonviolence.

Gandhi was distressed and pointed out that there was
hardly any contradiction involved, since the contribution to
the war effort was intended to hasten the conclusion of the
war and was, accordingly, a natural corollary to nonviolence.
He did not--perhaps could not--realize that the government
he was dealing with was always suspicious of him and doubted
his sincerity. Wavell's own account of Gandhi read: "Gandhi
is a remarkable old man certainly, and the most formidable of
the three opponents who have detached portions of the British Empire in recent years: Zaghlul and de Valera being
the other two, but he is a very tough politician and not a
saint."127

Gandhi offered his services for ending the war to Winston Churchill too, imploring him "to trust and use me for
the sake of your people and mine, and through them, those
of the world."128 But Churchill, true to his character and
still nursing his blindspot for India, could not overcome his
arrogance to oblige what he considered to be his biggest opponent, whom he had contemptuously described as the "half-
naked Fakir" following the Irwin-Gandhi agreement in 1930.
Accordingly, beyond a simple acknowledgment of Gandhi's
letter, there was no response from him.

The government of India, vide the viceroy's letter of
August 15, 1944, rudely brushed aside Gandhi's offer, which

pained him to no end. He maintained that his proposals had been

> presented on behalf of all the exploited nations and races of the earth. It is a great pity that the Lords and the Commons turned down my offer.
>
> The Allies will have their victory, but the exploited races will not feel the glow of it. They will know that the seeds of another and deadlier war will be sown by that victory.
>
> I ask myself the question, "Must rivers of blood flow from such an empty victory?"[129]

Gandhi felt depressed and conveyed his anguish to a Quaker friend in England, Agatha Harrison: "Everything I do turns to dust. It must be so, so long as I am 'Untrustworthy.'" But he would not give up; nothing that the British or their government of India did could cause him disappointment, so predictable was their attitude. So, he added in his letter to Miss Harrison, "if I represent the truth and if I do as God bids me, I know that the wall of distortion and suspicion will topple."

Resigned to the insensitivity and deep suspicion of the rulers of India, recovering their arrogance with the scent of eventual victory in the war, Mohandas Karamchand Gandhi on this note of faith and hope turned to his Constructive Program.

NOTES

1. For instance, as early as 1900, even while he was helping the British war effort, Gandhi felt that the Indians would continue to be treated as "social lepers" by the British. See The Collected Works of Mahatma Gandhi, vol. 3 (New Delhi: Government of India, Publications Division), p. 168. After Gandhi's experience of the Zulu uprising, he frequently dwelt in his writings upon the utterly selfish motives that led the British to occupy India. See ibid., vol. 8, p. 246; vol. 10, p. 22. Hind Swaraj is but a discovery of modern civilization, of which Britain, he maintained, was the true embodiment; thus, "the British Government in India constitutes a struggle between the modern civilisation, which is the Kingdom of Satan. . . ." See Mahatma Gandhi, Hind Swaraj (Ahmedabad: Navajivan, 1962), p. 189.

2. Collected Works, vol. 23, p. 116; also Gandhi's own statement, "in 1920, I became a rebel," ibid., vol. 48, p. 189; H.S.L. Polak, H.N. Brailsford, and Lord Pethick-Law-

rence, Mahatma Gandhi (London: Odhams Press, 1949), p. 126; C.F. Andrews, Mahatma Gandhi's Ideas (London: Allen & Unwin, 1949), pp. 230-231; B.R. Nanda, Mahatma Gandhi, A Biography (London: Allen & Unwin, 1959), pp. 191-199.

3. Harijan, vol. 4 (New York: Garland, 1973), p. 284.

4. Collected Works, vol. 21, pp. 438-439.

5. Ibid., vol. 20, p. 157.

6. Ibid., vol. 36, p. 86.

7. Ibid., pp. 108-109.

8. Ibid., vol. 63, p. 374.

9. Paul F. Power, Gandhi on World Affairs (Bombay: Perennial Press, 1961), p. 50.

10. Collected Works, vol. 37, pp. 270-271.

11. Reinhold Niebuhr, Moral Man and Immoral Society (New York: Charles Scribner, 1932), pp. 243-244.

12. Collected Works, vol. 40, p. 365.

13. Ibid., vol. 20, p. 109.

14. Ibid., vol. 16, p. 553; vol. 48, p. 189.

15. Ibid., vol. 36, p. 86; vol. 29, p. 64; vol. 30, p. 461.

16. Harijan, vol. 6, p. 290. According to Gandhi, "War is one powerful means, among others. But if it is a powerful means, it is also an evil one." Collected Works, vol. 15, p. 52; vol. 37, p. 271; vol. 39, p. 423.

17. Harijan, vol. 7, p. 265.

18. Ibid., pp. 265, 288.

19. Collected Works, vol. 41, p. 310; vol. 30, p. 461; vol. 40, p. 167.

20. Harijan, vol. 7, p. 436.

21. Collected Works, vol. 22, pp. 204-205.

22. Harijan, vol. 8, p. 302; vol. 7, p. 279.

23. Ibid., vol. 9, p. 60.

24. Collected Works, vol. 34, pp. 317-318; vol. 16, pp. 552-556.

25. Ibid., vol. 38, pp. 209-210.

26. Speeches and Writings of Mahatma Gandhi (Madras: G. A. Natesen, 1933), p. 419.

27. D.G. Tendulkar, Mahatma, vol. 6 (New Delhi: Government of India, Publications Division, 1960), p. 241.

28. Collected Works, vol. 23, p. 241.

29. Ibid., vol. 30, p. 461.

30. Ibid., vol. 48, pp. 409-410.

31. Ibid., vol. 26, pp. 126-129; vol. 50, p. 248.

32. Ibid., vol. 17, p. 489.

33. Harijan, vol. 9, p. 40.

34. Collected Works, vol. 25, p. 222.

35. Tendulkar, Mahatma, vol. 6, pp. 265-266.

36. Harijan, vol. 10, p. 212.

37. Ibid., vol. 6, p. 290.

38. Collected Works, vol. 23, p. 241.

39. Ibid., vol. 15, p. 406; vol. 32, pp. 401-402; vol. 48, p. 411.

40. Ibid., vol. 15, p. 487.

41. Hind Swaraj, pp. 28-29.

42. Harijan, vol. 9, p. 129.

43. Collected Works, vol. 40, p. 365.

44. Harijan, vol. 8, p. 129.

45. Ibid., vol. 6, p. 352.

46. Ibid., p. 290.

47. Ibid., vol. 12, p. 286; Collected Works, vol. 48, pp. 429-430.

48. Harijan, vol. 6, p. 290.

49. Tendulkar, Mahatma, vol. 6, p. 97.

50. Collected Works, vol. 25, p. 369.

51. Harijan, vol. 7, p. 8.

52. Ibid., vol. 9, p. 188.

53. C.W. De Kiewiet, A History of South Africa--Social and Economic (London: Oxford University Press, 1941), pp. 115-140.

54. Pyarelal, Mahatma Gandhi, The Last Phase, vol. 2 (Ahmedabad: Navajivan, 1958), p. 138.

55. Powers, World Affairs, p. 60.

56. Collected Works, vol. 41, p. 310; vol. 21, p. 354.

57. Bimal Prasad, Origins of Indian Foreign Policy, Indian National Congress and World Affairs, 1885-1947 (Calcutta: Bookland, 1960), pp. 138-143.

58. Collected Works, vol. 36, p. 86; vol. 37, pp. 270-271; vol. 39, p. 423; Autobiography, pp. 255-258.

59. Harijan, vol. 7, p. 265.

60. Ibid., p. 248.

61. Ibid., pp. 265, 272, 309.

62. Ibid., pp. 167-168, 279.

63. Ibid., p. 249.

64. Letter dated October 20, 1939, AICC File no. 5/1939-40, Nehru Memorial Museum Library, New Delhi. In part, the letter read: "In my opinion, the acceptance of the advice of Mahatma Gandhi as to unconditional offer of assistance to Britain in prosecution of War would have very much strength-

ened her claim to complete independence after the termination of the War."

65. Collected Works, vol. 15, p. 29.

66. Gandhi's acknowledgment dated August 2, 1941, of the invitation to attend such a conference says: "It does not attract me for the simple reason that I do not trust the British Government to do the right thing. Moreover, participation in conferences invited by the Government presupposes aid in war effort by the participants. I cannot, I will not." Gandhi Papers (New Delhi: Gandhi National Library and Museum).

67. Abul Kalam Azad, India Wins Independence (Bombay: Orient Longmans, 1959), p. 34.

68. Collected Works, vol. 48, p. 411.

69. Gandhi Papers, December 1939.

70. Harijan, vol. 7, p. 265.

71. Ibid., p. 272.

72. Letter dated October 31, 1939, from S.N. Madhok to the Congress president, AICC File no. 5/1939-40, N.M.M. Library.

73. Tendulkar, Mahatma, vol. 5, pp. 253-255; vol. 8, p. 148.

74. Ibid., pp. 208, 286.

75. Azad, India Wins Independence, p. 33.

76. Harijan, vol. 8, pp. 180, 269.

77. Ibid., pp. 185, 201.

78. Gandhi Papers, August 1940.

79. Ibid., Gandhi to Carl Heath, October 5, 1940.

80. AICC File no. G-7/1939-40, N.M.M. Library.

81. Tendulkar, Mahatma, vol. 6, pp. 4, 6, 14.

82. Ibid., pp. 33-34.

83. AICC File no. P-I (part 2) 1942, N.M.M. Library.

84. Tendulkar, Mahatma, vol. 6, pp. 36-37.

85. AICC File no. G-16 (1942-46), N.M.M. Library.

86. Nicholas Mansergh, ed., The Transfer of Power 1942-7, vol. 1 (London: Her Majesty's Stationery Office, 1970), p. 121; also, pp. 110-112.

87. Tendulkar, Mahatma, vol. 6, p. 62.

88. Louis Fischer, The Life of Mahatma Gandhi (Stuttgart: Tauchnitz, 1953), p. 310.

89. Letter dated April 22, 1942, Gandhi Papers.

90. Tendulkar, Mahatma, vol. 6, pp. 114, 171.

91. Harijan, vol. 9, pp. 128, 156, 172.

92. Gandhi to Rajagopalachari, July 5, 1942, Gandhi Papers.

93. J.B. Kripalani, Gandhi, His Life and Thought (New Delhi: Government of India, Publications Division, 1970), p. 200.

94. Jawaharlal Nehru, The Discovery of India (Calcutta: Signet Press, 1969 reprint), p. 473.

95. Azad, India Wins Independence, p. 76.

96. Harijan, vol. 7, p. 288.

97. Tendulkar, Mahatma, vol. 6, p. 27.

98. Ibid., p. 266.

99. Harijan, vol. 9, p. 168.

100. Mansergh, Transfer of Power, vol. 2, pp. 187-190.

101. Harijan, vol. 9, pp. 187, 228, 252, 264.

102. Pyarelal, Last Phase, vol. 1, pp. 10-11.

103. Gandhi's Correspondence with the Government 1942-44 (Ahmedabad: Navajivan, 1957), pp. 18-19, 26-28.

104. Mansergh, Transfer of Power, vol. 2, pp. 682, 684-686, 691, 730.

105. Penderel Moon, Wavell, The Viceroy's Journal (London: Oxford University Press, 1973), p. 461.

106. AICC File no. G-7/1939-40, N.M.M. Library.

107. Tendulkar, Mahatma, vol. 5, p. 221.

108. Gandhi Papers, January 1940.

109. Harijan, vol. 8, p. 208; vol. 9, p. 197.

110. Ibid., vol. 9, p. 258; vol. 10, p. 78.

111. Ibid., vol. 9, pp. 112, 232, 258.

112. Winston Churchill, The Second World War, vol. 2 (London: Cassell, 1949), pp. 355-359.

113. Gandhi's Correspondence, pp. 247-251.

114. Harijan, vol. 9, pp. 12, 48. An interesting aspect of the Gandhian approach is reflected in the following correspondence:

Letter to Gandhi dated October 29, 1941

> . . . Your recent permission to the [All India Spinning Association] to sell their blankets to the war department has however roused me to revise my opinion. Just as blankets do not play any part in murdering so do the services in the [Post and Telegraph]. There appears a very close analogy between the two.
>
> Even if I join as Technical Officer I shall not be taking direct part in violence. I know you are very busy these days to give attention to these petty details of an individual. But as you are the only Supreme Judge on these matters, please intimate your 'yes' or 'no' to me. You may explain it later on when you publish Harijan again.
>
> Yours Reverently
>
> Signed   Sant Singh
>   Sub-Post Master, Kalabagh
>   Mianwali, Punjab

Gandhi's reply dated November 3, 1941:

Dear Sant Singh

There is no analogy between my selling blankets to a murderer and your offering services on the field. Selling blankets to [illegible] I may sell them to whosoever requires them. To offer your services on the field is no part of your duty. My answer therefore is NO.

Signed M K Gandhi

115. Harijan, vol. 9, p. 136.

116. Ibid.

117. See, for instance, Harijan of April 26, May 24, June 7, 14, 21, 28, and July 5, 1942.

118. Harijan, vol. 9, pp. 137, 186.

119. Ibid.

120. Ibid., p. 188.

121. Ibid., pp. 128, 205.

122. Collected Works, vol. 40, p. 167.

123. Pyarelal, Last Phase, vol. 1, pp. 26-35.

124. Harijan, vol. 9, p. 212.

125. Tendulkar, Mahatma, vol. 6, p. 262.

126. Mahatma Gandhi's Correspondence with the Government 1944-47 (Ahmedabad: Navajivan, 1959), p. 6.

127. Moon, Wavell, The Viceroy's Journal, p. 236.

128. Gandhi's Correspondence 1944-47, pp. 7, 11, 15.

129. Tendulkar, Mahatma, vol. 6, pp. 265-266; Gandhi's Correspondence 1944-47, pp. 33-34.

5
GANDHI'S ENDURING POSITIVE STANCE

The panacea of nonviolence that he was perfecting and prescribed, Gandhi was convinced, had universal applicability. Some feel, however, that his view of the world was unrealistic and limited, for his knowledge and interest in international affairs was somewhat peripheral.[1] As a result, some sought to create the impression that, since Gandhi did not know much about world affairs, he did not quite understand the nature of fascism-nazism, nor did he appreciate the urgency of the threat that these twin evils posed to humanity. Nothing could be farther from the truth, nor more mischievous.

Apart from the substantial evidence that Gandhi followed the developments in the world closely, one cannot overlook that his perception of the machinations of international politics took shape as the direct result of life in the British Empire, particularly his native India. And what was this world he did not have knowledge about? A handful of European powers owned the bulk of the globe as colonies, with the Americas in "splended" isolation; the Soviet Union was callously consigned to the status of a pariah, and the Chinese were struggling to resist the encroachment of European and neighboring predators. It was the empires of the day that ran the affairs of the world, and the biggest of these conglomerates he knew firsthand. If you knew life under one empire, you knew all there was to know about the world that mattered, so to speak. The basis and nature of imperialism-colonialism, the quality of life under foreign rule, and the attitudes, interests, and style of the alien rulers were the same in all empires. Marginal differences of degree may be there, but never those of kind or quality.

Gandhi had experienced it all firsthand, and the condition of India roused the nationalist in him. That of the southern

African segment of the British Empire impelled him to experiment with truth for a good two decades. The result was conclusions and techniques that have international significance.[2]

Fascism-nazism were phenomena of the times, and the nations of Europe themselves had contributed toward these in no small measure. They had to be combated in the totality of their context and not in isolation from the overall situation. Waging war against fascism while continuing the exploitation of the empires was an indefensible contradiction, which Gandhi found unacceptable.

Accordingly, he had a basic difference of approach with the Allies in their war against fascism-nazism. The Allies thought of these systems entirely in geographical terms and sought to contain them by a "horizontal" strategy founded upon national boundaries. Gandhi, in contrast, saw fascism purely in ideological terms and pleaded for a "vertical" strategy to counter it. The fight against fascism, in his view, had to be waged wherever such tendency raised its ugly head.

The reason why the gains of strife, and war, were ever so shortlived, Gandhi maintained, was that men had always fought in terms of geography; and geographical divisions and factors, all know, are seldom permanent or reliable. Lasting gains like the establishment of durable peace, he said, could only be secured if the warring nations were to appreciate properly that advances in science and technology in modern times had rendered geographical factors quite irrelevant and insignificant. In any case, national boundaries almost invariably had been laid down rather arbitrarily, and by people: "God never made these frontiers," he rightly asserted.[3]

Gandhi's method to counter Nazi and Fascist aggression was through nonviolent defense and resistance. Long before World War II, Gandhi had maintained:

> For protecting others it is not necessary to possess the strength of the sword. The age of such strength has passed, or is passing. The world has had plenty of experience of the sword and has had enough of it. Even the West seems to have tired of it.
>
> He who protects others by killing the enemy is no Kshatriya [popularly described as the warrior caste among Hindus]; he is a true Kshatriya who protects others by laying down his own life. He is no brave man who runs away from danger; one who stands firm chest forward, and bears blows without striking back is a Kshatriya.[4]

He had noticed the effectiveness of nonviolent defense: "The Dukhobours, in Russia, met with no defeat. When it became impossible for them to continue to live in Russia, they left the country rather than submit to its oppressive rulers. To-day they live as highly respected community in Canada."[5]

Nonviolent noncooperation, which required no physical force but only self-sacrifice, according to Gandhi, could end only in success--there could be no defeat. For him the self-suffering bravely borne by the Boer women in 1899 had been exemplary, for they

> understood that their religion required them to suffer in order to preserve their independence, and therefore patiently and cheerfully endured all hardships.
>
> Lord Kitchener left no stone unturned in order to break their spirit. He confined them in separate concentration camps, where they underwent indescribable sufferings. They starved, they suffered biting cold and scorching heat. Sometimes a soldier intoxicated with liquor or maddened by passion might even assault these unprotected women. Still the brave Boer women did not flinch.
>
> And at last, King Edward wrote to Lord Kitchener saying that he could not tolerate it, and that if it was the only means of reducing the Boers to submission, he would prefer any sort of peace to continuing war in that fashion, and asking the General to bring the war to a speedy end.[6]

It was the self-suffering borne by these brave women that, in Gandhi's opinion, made the English relent. He similarly advocated erecting a human wall against foreign aggression and thus stop the rivers of blood flowing in World War II. When the Japanese were advancing upon India, he actually planned to send his disciple Mirabehn (formerly Madeline Slade) to Orissa for mobilizing such a human wall by instilling in them the invincible strength of nonviolent resistance.[7]

The idea of nonviolent resistance as a universal means of conflict resolution, applicable equally to the individual and the collectivity, national as well as international conflicts, is evident in Gandhi's Hind Swaraj. But he does not seem to have specifically articulated his views on the technique in relation to the defense of a nation in face of external aggression until he was moved to do so by the events preceding World War II. In his keenness first to avoid war and, after if it had started, to end it quickly, he not only desired the Congress not to involve itself in it, but made frantic appeals to the leaders of the nations involved in the war also to take recourse to nonviolent defense. Obviously, Gandhi had by now worked out a rough scheme of nonviolent defense against the foreign aggressor, following his revised stance on war in general.

The first blast of war came in the early 1930s when the Japanese invaded Manchuria. Gandhi considered Japan's aggression and devastation of China utterly unwarranted, unprovoked, and deplorable. Japan's adventure was quickly followed by Fascist Italy's attack on Abyssinia. Mussolini's brazenness stirred Gandhi even more to take significant interest in world affairs and comment on the international crisis of the day, even though his preoccupations with the freedom struggle at home were overwhelming. How could he possibly remain ignorant and aloof when humanity was on the very verge of annihilation and the agony of suffering was mounting precipitously the world over? Kind and sensitive as he was, Gandhi naturally reacted to the happenings in world politics in the columns of Harijan, from which we shall be quoting extensively in the next few pages.

With perfect consistency, Gandhi prescribed the technique of nonviolent defense to meet aggression. The Italian invasion of Abyssinia, incidentally, on Gandhi's birthday in 1935, gave him a strong enough indication of the disastrous course the great nations were setting themselves on. In such a situation, he saw his task to be to limit, as much as possible, the destruction and the misery war inevitably unleashes. So he advised the Abysinnians not to appeal to the League of Nations, or any other power, to intervene militarily on their side, but instead to resort to self-defense through the technique of nonviolent noncooperation:

> If Abyssinia were non-violent, she would have no arms, would want none. . . . Italy would find nothing to conquer if Abyssinians would not offer armed resistance, nor would they give cooperation willing or forced.
> Italian occupation in that case would mean that of land without its people. That, however, is not Italy's exact object. She seeks submission of the people of the beautiful land.[8]

In 1938 he reiterated that had the Abysinnians adopted "the attitude of non-violence of the strong, that is, the non-violence which breaks to pieces but never bends, Mussolini would have had no interest in Abyssinia. . . . He would certainly have been obliged to retire."[9]

Similarly, holding that the Japanese action in China was absolutely indefensible, Gandhi wished that China could respond to the challenge of aggression with nonviolence, even

die without using violence against the aggressors. He hoped
that the Chinese would rise above the morality of Japan and
urged that China should not offer armed resistance. For if
it did, "the victory of China will not mean a new hope for
the world . . . China will then be a multiple edition of Japan." China's victory by arms would only mean one more mil-
itarist nation. According to Gandhi, it was unbecoming for a
nation of 400 million, a nation as cultured as China, to repel
the Japanese invasion by using Japan's own violent method.
If, on the other hand, the Chinese resisted the Japanese
non violently, "there would be no use left" for the latter's
deadly weapons of destruction. The Chinese would simply say
to Japan: "Bring all your machinery, we present half of our
population to you. But the remaining 200 millions won't bend
their knee to you. . . . If China did that, Japan would become China's slave." In support, he quoted Shelley's The
Masque of Anarchy:

> Ye are many, they are few
> Stand ye calm and resolute,
> like a forest close and mute,
> With folded arms and looks which are
> weapons of unvanquished war.

The Arab-Jew dispute over Palestine exacerbated in the
late 1930s. The Jewish claim for a homeland in Palestine did
not particularly appeal to Gandhi, for he felt quite clearly
that Jews, like any other people in the world, should make
the country of their birth and domicile their home. In his
considered opinion, it was precisely this kind of claim that
regrettably emboldened Nazi Germany to persecute and expel
Jews. He absolutely loathed Hitler's persecution of the Jews:
in the name of religion, purity of race, and so on, Hitler
was committing a crime against humanity that, according to
Gandhi, was unparalleled in history. Thus he wanted the
great powers to restrain Hitler from his mad course. In fact,
Gandhi felt so strongly on the issue that were he a believer
in the method of war, and if a war had to be waged for the
purpose, he was willing to consider such a war fully justified.
"But I do not believe in any war," he maintained, "a discussion of the pros and cons of such a war is, therefore, outside my horizon or province." Yet Hitler must be resisted.
As Gandhi contemplated appealing to the great powers to
save the Jews, those very powers knuckled under and compromised to appease Hitler. Gandhi was amazed:

> If there can be no war against Germany, even
> for such a crime as is being committed against
> the Jews, surely there can be no alliance with
> Germany. How can there be an alliance between

a nation which claims to stand for justice and
democracy and one which is the declared enemy
of both? Or is England drifting towards armed
dictatorship and all that it means?

He advised the Jews, however, to offer satyagraha to
Hitler. For that alone could provide them with inner strength
and spare them from such treachery. Outside sympathies
could hardly be of any avail to them. "Indeed, even if Britain, France and America were to declare hostilities against
Germany," proclaimed Gandhi, "they can bring no inner joy,
no inner strength. The calculated violence of Hitler may even
result in a general massacre of the Jews by way of his first
answer to the declaration of such hostilities." And he was
not proven entirely wrong!
The Munich Pact distressed Gandhi greatly, and he
called it a betrayal of Czechoslovakia by Britain and France.
On Hitler's invasion of Czechoslovakia, Gandhi observed:

> The Czechs could not have done anything else
> when they found themselves deserted by their
> two powerful allies.
> And yet I have the hardihood to say that if
> they had known the use of non-violence as a
> weapon for the defence of national honour, they
> would have faced the whole might of Germany
> with that of Italy thrown in. . . . And to save
> their honour they would have died to a man
> without shedding the blood of the robber.
> I must refuse to think that such heroism, or
> call it restraint, is beyond human nature. . . .
> I suggest that if it is brave, as it is, to
> die to a man fighting against odds, it is braver
> still to refuse to fight and yet to refuse to
> yield to the usurper.

In answer to the objection that his noble technique would
not be effective on a soulless person like Hitler, Gandhi, in
an article, "If I Were A Czech," held:

> History has no record of a nation having
> adopted non-violent resistance. If Hitler is
> unaffected by my suffering, it does not
> matter. For I shall have lost nothing of worth.
> My honour is the only thing worth preserving.
> That is independent of Hitler's pity.
> But as a believer in non-violence, I may
> not limit its possibilities. Hitherto he and his
> likes have built upon the invariable experi-
> ence that men yield to force. Unarmed men,

women and children offering non-violent resistance without any bitterness in them will be a
novel experience for them.

Who can dare say that it is not in their nature to respond to the higher and finer forces?
They have the same soul that I have.

Gandhi was inclined to think that, in the face of nonviolent
resistance by the Czech people, the invading troops would
refrain from violence, though they would occupy the territory and take possession of what they wanted nonetheless.
Even such a situation was to be preferred to violent resistance, and so, in the event of the invaders occupying mines,
factories, and all the natural resources belonging to the
Czechs, Gandhi envisaged the following consequences:

(1) The Czechs may be annihilated for disobedience to orders. That would be a glorious victory
for the Czechs and the beginning of the fall of
Germany.
(2) The Czechs might become demoralised in the
presence of overwhelming force. . . . But if
demoralization does take place, it would not be
on account of non-violence, but . . . due to absence or inadequacy of non-violence. . . .
(3) Germany might use her new possessions for
occupation by her surplus population. This,
again, could not be avoided by offering violent
resistance, for we have assumed that violent resistance is out of the question.

Thus non-violent resistance is the best method under all conceivable circumstances.

In a letter dated July 23, 1939, Gandhi implored Hitler
to desist from plunging the world into war and tried to persuade him to the path of nonviolence instead:

Friends have been urging me to write to you
for the sake of humanity. But I have resisted
their request because of the feeling that any
letter from me would be an impertinence.

Something tells me that I must not calculate
and that I must make my appeal for whatever
it may be worth.

It is quite clear that you are today the one
person in the world who can prevent a war
which may reduce humanity to the savage state.
Must you pay the price for an object, however
worthy it may appear to you to be? . . .

> Anyway I anticipate your forgiveness, if
> I have erred in writing to you.

The letter failed to have the desired effect, of course. In the words of Shridharani, "Perhaps Gandhi was a fool, but a God's fool, who regarded no one as utterly lost because man is a reflection of the divine and as such susceptible to regeneration under all circumstances."[10]

Gandhi's faith in nonviolent resistance was rooted in his conviction that most people, however prone to violence, were not incurably addicted to it. Man as an animal may be violent, but in spirit he was nonviolent, and the moment he awakens to the spirit he cannot remain violent.[11] The doctrine of ahimsa, apparently, calls upon the human beings to regard themselves not merely as superior animals, but to remind themselves of their moral and spiritual status and obligations as well. However imperfect this realization, held Gandhi, it still acts as a brake against violence and arouses the power of nonviolence, or love, latent in every human soul. Fundamentally, Gandhi believed in what he called the absolute oneness of God and man. The rays of the sun are a myriad through refraction, but they have the same source. Man is bestial in origin, but he is human precisely because he is potentially, and essentially, divine: we are thus born with brute strength, but we are born human to realize that God dwells in us. This, indeed, is the privilege of man, and it distinguishes him from the brute creation. Gandhi's invocation of ahimsa was ultimately an appeal to the conscience and the reason of the individual, an affirmation of the dignity of the human soul.

Hitler, as it turned out, was beyond redemption and pushed the world into war, when without much provocation he marched his armies on September 1, 1939, into Poland. In his dejection, Gandhi lamented, "it almost seems as if Herr Hitler knows no God but brute force, and as Mr. Chamberlain says he will listen to nothing else."[12] But a spurt of his faith in the redemptive nature of humanity impelled Gandhi to offer his services to negotiate with the German leader. "Hitler is not a bad man," he wrote to Viceroy Linlithgow, "If you call off [the war] today, he will follow suit. If you want to send me to Germany or anywhere else, I am at your disposal. You can also inform the Cabinet about this."[13]

In the same belief, Gandhi addressed his second, and open, letter to Hitler on December 24, 1941.[14] However, intelligence officials of the government of India for reasons best known to the government--possibly for false notions of "security"--never allowed the letter to reach the addressee. The letter, meant also for Mussolini, disapproved of Hitler's actions in no uncertain terms, elucidated India's nonviolent struggle for independence, and urged immediate restoration of peace. "That I address you as a friend," wrote Gandhi, "is no formality, I own no foes. My business in life for the past thirty-three years has been to enlist the friendship of the whole of humanity by befriending mankind, irrespective of race, colour or creed." Hoping that he would have the time and the desire to know how the rest of mankind viewed his actions, Gandhi assured Hitler that there was hardly any doubt about the latter's bravery or devotion to his fatherland.

> But your writings and pronouncements and
> those of your friends and admirers leave no
> room for doubt that many of your acts are
> monstrous and unbecoming of human dignity
> especially in the estimation of men like me who
> believe in universal friendliness. Such are
> your humiliation of Czechoslovakia, the rape
> of Poland and the swallowing of Denmark. I
> am aware that your view of life regards such
> spoliations as virtuous acts. But we have
> been taught from childhood to regard them as
> acts degrading to humanity. Hence we cannot
> possibly wish success to your arms.

Gandhi pointed out that the position and the struggle of Indian nationalists were quite unique.

> We resist the British imperialism no less than
> Nazism. If there is a difference, it is in de-
> gree. One-fifth of the human race has been
> brought under the British heels by means that
> will not bear scrutiny. Our resistance to it
> does not mean harm to the British people. We
> seek to convert them, not to defeat them on
> the battlefield.
> Ours is an unarmed revolt against the Brit-
> ish rule. But whether we convert them or not,
> we are determined to make their rule impossible
> by non-violent non-cooperation.

Gandhi's method had indeed a fair measure of success, and Hitler was possibly aware of that, albeit without acknowledging it. So Gandhi went on to stress the merit and the strength of his technique that the Indian struggle for independence had followed.

> It is a method in its nature undefeatable. It is
> based upon the knowledge that no spoliator
> can compass his end without a certain degree
> of co-operation, willing or compulsory; they can

have the former only by complete destruction
of every Indian--man, woman or child. That all
may not rise to that degree of heroism and
that a fair amount of frightfulness can bend
the back of revolt is true; but the argument
would be beside the point. For, if a fair number of men and women can be found in India
who would be prepared, without any ill-will
against the spoliators, to lay down their lives
rather than bend the knee to them, they will
have shown the way to freedom from the tyranny of violence. I ask you to believe me when I
say that you will find an unexpected number of
such men and women in India. They have been
having that training for the past twenty years.
. . .
    In non-violent technique, as I have said,
there is no such thing as defeat. It is all "do
or die", without killing or hurting. It can be
used practically without money and obviously
without the aid of the science of destruction
which you have brought to such perfection."

Gandhi marvelled that Hitler did not see that violence and the
force of arms were nobody's monopoly.

If not the British, then some other power will
certainly improve upon your method and beat
you with your own weapon. You are leaving no
legacy to your people of which they would feel
proud. They cannot take pride in a recital of
cruel deeds, however skilfully planned.
    I therefore appeal to you in the name of hu-
manity to stop the war. . . .

Stressing that the peoples of Europe yearned for peace, Gandhi continued in his letter to Hitler: "we have suspended
even our own peaceful struggle. It is not too much to ask you
to make an effort for peace during a time which may mean
nothing to you personally, but which must mean much to the
millions of Europeans whose dumb cry for peace I hear, for
my ears are tuned to hearing the dumb millions."
    Gandhi concluded his letter to Hitler, which the govern-
ment of India prevented from reaching him, with the observation: "I had intended to address a joint appeal to you and
Signor Mussolini, whom I had the privilege of meeting when I
was in Rome during my visit to England as a delegate to the
Round Table Conference. I hope that he will take this as addressed to him also with the necessary changes."

The dictators, true to their character, contemptuously disregarded Gandhi's views and appeals. But the suffering Poles looked up to him for counsel and consolation. Paderewski, former president of the Polish Republic (1920-21) and now the leader of the Polish government in exile, urged Gandhi in a cable to send a word of sympathy and encouragement to the Poles whose innumerable innocent women and children were massacred daily.[15]

Gandhi's whole heart was with the brave Poles "in the unequal struggle in which they [were] engaged for the sake of saving their freedom," but his misfortune was that he was in no position to render them help. However, he sent his "heart-felt prayer for the early termination of their fearful trial and for the grant of the required strength to bear the suffering whose very contemplation makes one shudder." Gandhi held the cause of the Polish people to be just "and their victory certain. For God is always the upholder of Justice." He declared: "If Poland has that measure of uttermost bravery and an equal measure of selflessness, history will forget that she defended herself with violence. Her violence will be counted almost as non-violence [emphasis added]."

Now this might seem an inconsistency, if not a contradiction, in Gandhi's position, but we do not think so. In the first place, he allows the use of violence only for the purpose of self-defense, when (a) nothing else is available and the strength of nonviolent resistance is not known to them, and (b) unprepared as the Poles were, the tendency in their panic was to flee in the face of overwhelming brute force. It was better to take a stand violently than to slink away in cowardice. Secondly, he saw that the duty of the people of Poland clearly lay in defending their country; if they discharged it with the help of pitchforks and whatever weapons were available at the moment, they should use these even though death was certain. The method and the path of nonviolence were infinitely and always superior, of course, insisted Gandhi. However, those who are unable to offer nonviolent resistance under exceptional circumstances forced upon them may offer violent resistance out of a sense of duty rather than cowardly surrendering to the aggressor.

Unmistakably, Gandhi makes a positive distinction between the use of violence in self-defense, to protect the honor and life of helpless women, children, and the aged, and its exercise for aggressive purposes. The seeming inconsistency, in fact, was pointed out to Gandhi when he was asked about his objection to the Wardha Resolution of October 23, 1939, in accordance with which the Congress ministries in the provinces had resigned by way of protest against British high-handedness. "If Polish resistance to the German invasion was almost non-violent and you would reconcile yourself with it why do you object to the Wardha Resolution?"

However, he saw no inconsistency in accepting violent defense of Poland and his commitment to nonviolence:

> Surely there is no analogy between the two cases. If a man fights with his sword single-handed against a horde of dacoits armed to the teeth, I should say he is fighting almost non-violently. Haven't I said to our women that, if in defence of their honour they used nails and teeth and even a dagger, I should regard their conduct non-violent? She does not know the distinction between Himsa and Ahimsa. She acts spontaneously.
>
> Supposing a mouse in fighting a cat tried to resist the cat with his sharp teeth would you call that mouse violent?
>
> In the same way, for the Poles to stand valiantly against the German hordes vastly superior in numbers, military equipment and strength, was almost non-violence.
>
> I should not mind repeating the statement over and over again. You must give its full value to the word almost.
>
> But we are 400 millions here. If we were to organise a big army and prepare ourselves to fight foreign aggression, how could we by any stretch of imagination call ourselves almost non-violent, let alone non-violent? The Poles were unprepared for the way in which the enemy swooped upon them.

Reiterating his ideal position, however, he wondered: "Would Poland unarmed have fared worse if it had met the challenge of the combined forces with the resolution to face death without retaliation?"

Gandhi seriously doubted if Hitler would at all be able to digest as much power as he was seeking: eventually, he was certain, Hitler also would leave the world as did Alexander--emptyhanded. The reason for this was not far to seek: it would be impossible for Hitler, he believed, to sustain the burden of a mighty empire and hold all the conquered peoples in perpetual subjection. In his essay, "How to Combat Hitlerism," Gandhi argued, "All the blood that has been spilled by Hitler has added not a millionth part of an inch to the world's moral stature. . . . Hitlerism can never be defeated by counter-Hitlerism; it can only breed superior Hitlerism to nth degree. . . ." In comparison to superior violence, he considered the French surrender of 1940 far better:

> I think French statesmen have shown rare cour-
> age in bowing to the inevitable and refusing to
> be a party to senseless mutual slaughter. There
> can be no sense in France coming out victorious
> if the stake is in truth lost. The cause of liber-
> ty becomes a mockery, if the price to be paid
> is wholesale destruction of those who are to en-
> joy liberty. It then becomes an inglorious sati-
> ation of ambition.

Continuing the argument on the means to combat Hitlerism,
Gandhi pointed out that the situation in Europe would have
been different

> if the Czechs, the Poles, the Norwegians, the
> French and the English had all said to Hitler:
> you need not make your scientific preparations
> for destruction. We will meet your violence with
> non-violence. You will, therefore, be able to
> destroy our non-violent army without tanks,
> battle ships and airships.
>   It may be retorted that the only difference
> would be that Hitler would have got without
> fighting what he has gained after a bloody
> fight.
>   Exactly, the history of Europe would then
> have been written differently. Possession
> [might] have been then taken under non-violent
> resistance, as it has been taken now after per-
> petration of untold barbarities.
>   Under non-violence only those would have
> been killed who had trained themselves to be
> killed, if need be, but without killing anyone
> and without bearing malice toward anybody. I
> dare say that in that case Europe would have
> added several inches to its moral stature. And
> in the end I expect it is the moral worth that
> will count. All else is dross.

Gandhi knew very well that none of the European nations had
any faith in nonviolence as a creed. In that essay, he was
advising them to use nonviolence even as a policy as, indeed,
he had been urging the Congress to do.
  Gandhi kept up the refrain ceaselessly. In his July 1940
article, "To Every Briton," Gandhi implored support of his
path:

> I appeal to every Briton, wherever he may be
> now, to accept the method of non-violence

instead of that of war for the adjustment of
relations between nations. . . .

I venture to present you with a nobler and
braver way worthy of the bravest soldier. I
want you to fight Nazism without arms, or, if I
am to retain the military terminology, with nonviolent arms.

I would like you to lay down the arms you
have, as being useless for saving you or humanity. You will invite Herr Hitler and Signor Mussolini to take what they want of the countries
you call your possessions. Let them take possession of your beautiful island, with your many
beautiful buildings. You will give all these, but
neither your souls, nor your minds. . . . You
will allow yourself, man, woman and child to be
slaughtered, but you will refuse to owe allegiance to them.

The appeal, conveyed at Gandhi's request by the viceroy to
the British government, drew the typical response: "With
every appreciation of your motives [the British government]
do not feel that the policy which you advocate is one which
it is possible for them to consider, since in common with the
whole empire they are firmly resolved to prosecute the war
to a victorious conclusion." Gandhi had never questioned the
"firm resolve" or the "victorious conclusion." In fact, these
were the very objectives that had prompted the appeal. He
was suggesting a nobler and better way--one that avoided
slaughter and suffering--to achieve victory: "Peace has its
victories more glorious than those of war."

Gandhi was satisfied that his "non-violent method would
have meant no abject surrender." Actually "it would have
confounded all modern tactics of war, indeed, rendered them
of no use." He was more than convinced that Hitler could
"only be confounded by the adoption of the novel method of
fighting. At one single stroke he will find that all his tremendous armament has been put out of action." The reason is that
warriors live on their wars, whether offensive or defensive,
and suffer a collapse if they find that their warring capacity
or faculties are unwanted. It simply would not be possible
for warriors to devastate any territory indefinitely were there
to be no violent resistance to them. "Even if Herr Hitler was
so minded," said Gandhi, "he would not devastate 700,000
non-violent villages. He would himself become non-violent in
the process."

Gandhi cast the same theme in the context of the threatened Japanese invasion in his letter of June 14, 1942, to
Chiang Kai-Shaik (later he withheld publication of the letter
in Harijan against his own inclination in deference to a cable

from Chiang Kai-Shaik.[16] In the letter Gandhi explained that his appeal to the British to withdraw from India did not mean in any shape or form the weakening of India's defense against the Japanese or embarrassing the Chinese in their struggle:

> India must not submit to any aggressor or invader and must resist him. . . . A Japanese domination of either India or China would be equally injurious to the other country and to world peace. That domination must therefore be prevented and I should like India to play her natural and rightful part in this.[17]

In fact, Gandhi had already reconsidered his position in regard to the stationing of Allied troops on Indian soil, when the Japanese threat became manifestly ominous. In his letter of July 1, 1942, to President Roosevelt, he reaffirmed:

> I hate all war. If, therefore, I could persuade my countrymen, they would make a most effective and decisive contribution in favour of an honorable peace. But I know that all of us have not a living faith in non-violence.
> Under foreign rule, however, we can make no effective contribution of any kind in this war, except as helots.[18]

As a well-wisher of the Allies, Gandhi offered that if the Allies think it necessary they may keep their troops, at their own expense, in India, not for keeping internal order but for preventing Japanese aggression and defending China. So far as India was concerned, it must become free even as the United States and Great Britain were. The Allied troops, he made it quite clear, would remain in India under treaty with the free government that may be formed by the people of India without any outside interference, direct or indirect.

Gandhi spelled out the content and style of his nonviolent defense thus:

> Japan is knocking at our gates. What are we to do in a non-violent way?
> If we were a free country, things could be done non-violently to prevent the Japanese from entering the country.
> As it is, non-violent resistance could commence the moment they [the Japanese] effected a landing. Thus, non-violent resisters would refuse them any help, even water. For it is no part of their duty to help anyone steal their country. . . .

> Such resisters will calmly die wherever they
> are but will not bend their knee before the ag-
> gressor. . . .
> But if a Japanese had missed his way and
> lay dying of thirst and sought help as a human
> being, a non-violent resister, who may not re-
> gard anyone as his enemy, would give water to
> the thirsty one.
> The underlying belief in such non-violent
> resistance is that the aggressor will, in time,
> be mentally and even physically tired of killing
> non-violent resisters.[19]

By way of preparing his countrymen to offer noncooperation
to the aggressor, Gandhi enjoined them:

> Our attitude is that of complete non-co-opera-
> tion with Japanese army. Therefore we may not
> help them in any way nor may we profit by any
> dealings with them. . . .
> One thing [people] should never do--to
> yield willing submission to the Japanese. That
> will be a cowardly act, and unworthy of a free-
> dom-loving people. They must not escape from
> one fire only to fall into another and probably
> more terrible. . . .[20]

Accordingly, Indians were not to cooperate with the invaders in the construction of bridges, for example, and were
under no circumstances to accept Japanese currency; they
were to depend, primarily, on barter for dealing among themselves or make use of such British currency as they might
have. They could lend their cooperation to the invading Japanese, however, for the removal and burial of dead bodies.
Furthermore, in keeping with his fondness for villages,
Gandhi discussed the desirability of exodus of the people
from urban centers to villages; in the bombed cities those
not actively engaged in the task of defense were an unnecessary burden to the defenders and thus must be evacuated to
the villages.[21] In resisting the invader, however, a policy of
"scorched earth"--a favorite of the Soviets--was not to be
employed under any circumstances. That kind of policy, in
Gandhi's opinion, was clearly suicidal and unnecessary; he
saw "neither bravery nor sacrifice in destroying life or property" and believed it to be a self-defeating measure. Out of
sheer humanitarian considerations and, of course, due to the
fact that Gandhi never took anyone to be his enemy, he would
much rather save food and clothing for the enemy than destroy them altogether. Food and clothing--already in short
supply--deserved urgent consideration both from the war re-

sisters and the war managers, and in India, therefore, the policy of scorched earth would be wholly indefensible.

In his appeal "To Every Japanese," condemning their "unprovoked attack" and merciless devastation of China, Gandhi cautioned the Japanese against invading India: "I would ask you," he wrote, "to make no mistake about the fact that you will be sadly disillusioned if you believe that you will receive a willing welcome from India." In fact, they should bring the war to an end, for sheer humanity demanded and deserved it: "Ruthless warfare is nobody's monopoly. If not the Allies some other power will certainly improve upon your method and beat you with your own weapon. Even if you win you will leave no legacy to your people of which they would feel proud. . . ."22

Gandhi advised war resisters everywhere not to lose faith in nonviolence in the face of the mounting odds. On hearing that Dr. Maude Royden, a British pacifist, had revised her opinion and was losing faith in the ability of the European nations to use spiritual power, he became very disheartened. In a long article, "On Its Trial," in the Harijan of March 15, 1942, he advised Dr. Royden to cling to her faith in nonviolence and continue her antiwar effort: "There is no cause whatsoever," wrote Gandhi, "for despondency, much less for denial of one's faith at the crucial moment. Why should not British pacifists stand aside and remodel their life in its entirety? They might be unable to bring about peace outright, but they would lay a solid foundation for it."23 To the observation that nonviolence by the Jews against persecution by Hitler had been of no avail, Gandhi's answer was that their nonviolence was merely a passive resistance of the weak and helpless; they were violent at heart and nonviolent only in appearance. If the Jews "instead of being helplessly and of necessity non-violent, adopt active non-violence, i.e., fellow feeling for the Gentile Germans deliberately . . . this supreme act of theirs will be the greatest contribution and war will be then a thing of the past."24

Apart from expressing himself on war in the context of the prevailing situation in India or the world at the time, Gandhi also encountered hypothetical queries on external defense in general.

Thus, in answer to the problem of a neutral country through which an invader wishes to pass, Gandhi held:

> It would be cowardly of a neutral country if you allowed an army to devastate a neighboring country.
>
> But there are two ways in common between soldiers of war and soldiers of non-violence and if I had been a citizen of Switzerland and a President of the Federal state, what I would

have done would be to refuse passage to the
invading army by refusing all supplies.

Secondly, re-enacting a Thermopylae in
Switzerland you would have presented a living
wall of men and women and children, and inviting the invaders to walk over your corpses.

You may say that such a thing is beyond
human experience and endurance. I say it is
not so. It was quite possible. Last year in
Gujarat, women stood Lathi charges unflinch-
ingly and in Peshawar thousands stood hails of
bullets without resorting to violence. . . .

The army would be brutal enough to walk
over them, you may say. I would then say you
will still have done your duty by allowing your-
selves to be annihilated.

An army that dares to pass over the corpses
of innocent men and women would not be able to
repeat that experiment.[25]

Shridharni, raising the question as to how an army could
be stopped at the border by a nonviolent state, asked:

Suppose a free India adopts Satyagraha and not
violence as an instrument of national policy, how
would she defend herself against probable aggression by another state? What kind of resistance could and would be offered to invader?
What would India's actions be to meet the invading army at the frontier? Or would she withhold
all action until after the invader had taken over
the whole country?

Gandhi offered what he called a speculative answer in a signed
article in the Liberty of August 17, 1940, which indicated that
he was not sure if free India would adopt civil disobedience
and nonviolence as its formal policy. For the sake of the argument, he said, "let us suppose that such is the case . . .
that there is no Indian army, no defensive fortification, no
rifles, cannons, shells, airplanes, tanks. And let us suppose
that India stands entirely by herself and that the vast and
powerful armies of a modern edition of Nero descend upon her."
In that event the manner in which India would defend itself
would be to

Let the invader in without opposition. But they
would tell the invader . . . on the frontier that
the Indian people would refuse to co-operate in
any work, in any undertaking. They would refuse to obey orders despite all threats and . . .

punishment inflicted upon them. That is civil
disobedience. That is India's defence.

You may fancy that the hardened and ruthless invader would laugh at such measures. If
he had conquered armies who opposed him with
steel and cannons and warplanes, surely it
would be ridiculously easy for him to conquer
this unarmed army!

But India is a land of millions, and if they
stand idle the whole country stands idle. Nothing can be done with it; it is worthless. Civil
disobedience, the invader would soon find, is
a very powerful weapon indeed.

In another statement Gandhi reiterates his plea for satyagraha.

Trained in the art of non-violent resistance,
the Indian people would offer themselves unarmed as fodder for the aggressor's cannons.
They would tell the invader that they preferred
death to submission. These brave words have
been spoken in other lands; in India they would
be spoken with all their true meaning, and spok-
en in one great overwhelming voice of the masses.
By the million, India's people would offer their
breasts to the invader's bullets. And this would
be a terrifying spectacle and one of the highest
moral stature, ennobling those who took part in
it.

The basic point Gandhi made was that nobody in this world
was entirely without heart; everyone had a conscience, howsoever constricted or warped. And so he emphasized without
equivocation:

The underlying belief in [such a] philosophy of
defence is that even a modern Nero is not devoid of a heart. The spectacle--never seen before by him or his soldiers--of endless rows of
men and women simply dying, without violent
protest, must ultimately affect him. If it does
not affect Nero himself, it will affect his soldiery.

Men can slaughter one another for years in
the heat of battle, for then it seems to be a case
of kill or be killed. But if there is no danger of
being killed yourself by those you slay, you cannot go on killing defenceless and unprotesting
people endlessly. You must put down your gun
in self-disgust.

> Thus in the end the invader must be beaten
> by new weapons, peaceful weapons, the weapons
> of civil disobedience and non-violent resistance.
>   Practically speaking, there would be probably
> no greater loss in life than if forcible resistance
> was offered to the invader. How many men have
> been killed in Holland, Belgium, and France?
> Hundreds of thousands? Would the invading
> armies have shot down hundreds of thousands
> of men in cold blood if they had simply stood
> passively before them? I do not think so.[26]

Gandhi, therefore, sincerely and without the slightest
doubt ever, believed that nonviolent resistance was bound to
succeed, and even the best conceived and highly equipped
armed, violent, resistance was always doomed to bitter failure. And the obvious, but scarcely appreciated, reason for
this was that the aggressor also was a human being, and
with a heart.

But what was to be done in the event of aerial warfare,
for in that situation there was no personal contact between
the satyagrahis and the aggressor? Gandhi's answer was:
"Behind the death dealing bomb there is the human hand in
motion."[27] The satyagrahis would, accordingly, employ the
same measures at airfields where the planes eventually alight
--for invaders have to alight at some place, some time to engage and occupy the territory and to rule the people in it.

Margret Bourke White, a U.S. journalist, asked Gandhi
hours before his death, "How would you meet the Atom Bomb
. . . With non-violence?" and Gandhi replied:

> I will not go underground. I will not go into
> shelter. I will come out in the open and let the
> pilot see I have not a trace of evil against him.
> The pilot will not see our faces from his great
> height, I know. But that longing in our hearts
> --that he will not come to harm--would reach up
> to him and his eyes would be opened. . . .
> Thousands who were done to death in Hiroshima,
> if they had died with that prayerful action . . .
> died openly with that prayer in their hearts
> . . . their sacrifice would not have gone in vain
> and the war would not have ended so disgrace-
> fully as it had.[28]

Manifestly, his faith in the efficacy of nonviolence, and in
the pursuit of truth, was so completely unshakable that even
the atomic bomb failed to cast any doubts in his mind. "There
have been Cataclysmic changes in the world," he wrote,

So I still adhere to my faith in truth and non-violence? Has not the atom bomb exploded that faith?

Not only has it <u>not</u> done so but it has clearly demonstrated to <u>me</u> that the twins [truth and non-violence] constitute the mightiest force in the World. Before it the atom bomb is of no effect.

The two opposing forces are wholly different in kind, the one moral and spiritual, the other physical and material. The one is infinitely superior to the other which by its very nature has an end. The force of spirit is ever progressive and endless. Its full expression makes it unconquerable in the World.[29]

There was absolutely no doubt whatsoever in Gandhi's mind that nonviolence was "the only thing that the atom bomb could not destroy. I did not move a muscle when I first heard that the atom bomb had wiped out Hiroshima." He considered the use of the atomic bomb "as the most diabolical use of science." At the same time, however, it had also squashed the last iota of doubt that "Unless now the world adopts non-violence, it will spell certain suicide for mankind."[30]

If nonviolence alone could redeem humanity, was there any hope for those who had given all their active life to fighting? A senior officer of the INA asked: "How can one who has spent his whole life in fighting, take to <u>Ahimsa</u> with success? Are not the two incompatible?" Gandhi did not accept the proposition:

Badshah Khan [Abdul Gaffar Khan] is a Pathan. But today he has become a soldier of non-violence. Tolstoy, too, served in the army. Yet he became the high-priest of non-violence in Europe.

We have not yet realised the power that is non-violence. If the Government had not arrested me in 1942, I would have shown how to fight Japan by non-violence.[31]

Consistently, Gandhi was not inclined to favor the retention of an army in free India. "What has the army done for India?" he asked,

I fear that in no sense has it served India's interest. It has kept millions of inoffensive and disarmed people under subjection. It has impoverished them. . . .

> If free India has to sustain the present mili-
> tary expenditure, it will bring no relief to the
> famishing millions.[32]

Even after the attainment of independence, he wanted to see
the

> defenders of Kashmir dying at their post of
> duty without malice and without anger in their
> hearts against the assailants, and without the
> use of arms including even their fists. . . .
> Kashmir would then become a holy land
> shedding its fragrance not only throughout In-
> dia, but the world.[33]

Referring to the post-partition situation in the Indian subcontinent, Gandhi declared:

> India knows, the world should, that every
> ounce of my energy has been and is being de-
> voted to the definite avoidance of fratricide
> culminating in war.
> When a man vowed to non-violence as the
> law governing human beings dares to refer to
> war, he can only do it so as to strain every
> nerve to avoid it.
> Such is my fundamental position from which
> I hope never to swerve even to my dying day.[34]

Gandhi viewed the modern state as one based on force.
For such a state the offering of nonviolence was not possible.
But "a state," he maintained, "can be based on non-violence,
i.e., it can offer non-violent resistance against a world combination based on armed forces. Such a state was Asoka's
[and] the example can be repeated."[35] On the rare occasions
that he spoke, or wrote, about the nonviolent state, he
talked vaguely about eliminating the armed forces but retain-
ing the police as a regulating, supervising, and monitoring
agency of the government.

One can reasonably discern, however, what he has in
mind with regard to the nature and role of the state. His
choice of Emperor Ashoka's state by way of illustration is
significant. Ashoka's state assumed its nonviolent character
after the bloody battle of Kalinga, which took the lives of
nearly 100,000. Such a heavy price for victory weighed on
the young monarch's conscience and led him to embrace Buddhism which eventually became the state religion. It was
thus a state based more on a measure of cooperative participation than mere implicit consent or force. This state was
not devoid of all conflict; its strength lay in peaceful resolu-

tion of conflict rather than in the conflict propelling it to war. The dynamism of peaceful resolution of conflicts, and persuasion as the basis of consensus, alone were responsible for the growth of Ashoka's state.

In choosing the example of that state, Gandhi reveals his thinking thus. He points to the needlessness of force or coercion as the basis of the state, while, at the same time, duly accepting the social fact of conflict. Combine it with his other vital premises--swaraj and a high sense of duty on the part of the citizen, upon both of which he constantly insists--and you have a nonviolent state neither capable of nor in need of waging war. Obviously, the state here rests upon minimal violence required only for regulatory purposes rather than for domination or exploitation. The dynamics of such a situation effortlessly and automatically ensures an enduring peace.

Gandhi just could not bring himself to accepting the inevitability of war. His whole effort was geared toward not merely eliminating it but, in fact, to creating conditions of peace that naturally rule out the possibility of armed conflict among nations: hence the attempt to label his action-oriented nonviolent philosophy and approach as war without violence. A moral equivalent of war, in his thinking, was perhaps uncalled for. War had a particular connotation for him, as should be evident from the foregoing discussion. His long-term objective, it follows, was to induce a peaceful world order where nations could live together in harmony.

Gandhi thus campaigned unremittingly not so much for avoidance of war as, indeed, for the elimination of the causes of war. To all outward appearances, World War II ended in 1945, but for Gandhi it was still on, and there was no avoiding that reality: "cold" or otherwise, war continued unremittingly to engage and envelop nations in its destructive sweep; it could not be eliminated:

> Unless big nations shed their desire of exploitation and the spirit of violence of which war is a natural expression and atom-bomb its inevitable consequence, there is no hope for peace in the world.
> I tried to speak out during the War and was dubbed as Fifth-columnist for my pains.[36]

If Gandhi were not heard then, the postwar world--vitiated so thoroughly by the ever-expanding cold war--was in a still lower mood to pay any attention to him. The cacophony of the cold war and the macabre vision of a nuclear holocaust effectively drowned his firm but lonely voice.

Undeterred, Gandhi remained firm in his conviction and optimism that humanity had steadily been moving away from himsa and turning to ahimsa;[37] that people's hunger for the latter grew with every passing day. In his lifetime, humanity had experienced the saturation point of violence in the form of World War II, and he saw everyday a growing testimony of the fact that "Ahimsa was never before appreciated by mankind as it is today."[38]

The Young India of June 23, 1919, concluded:

> It may be long before the law of love will be
> recognised in international affairs. The machin-
> eries of Governments stand between and hide
> the hearts of one people from those of another.
> Yet . . . we could see how the world is moving
> steadily to realise that between nation and na-
> tion as between man and man, force has failed
> to solve problems [p. 50].

It was Gandhi speaking through the paper he was then editing.

Many years later, he still seemed confident that masses would refuse to bow to the moloch of war and would "rely upon their capacity for suffering to save [their] country's honour."[39]

NOTES

1. See, for instance, Robert Payne, The Life and Death of Mahatma Gandhi (London: Bodley Head, 1969), p. 485; H.A. Jack, ed., The Gandhi Reader (Bloomington: Indiana University Press, 1956), p. 332.

2. There may seem to be some parallel here between Gandhi and Mao, both makers of their respective nations whose ideas and approach have acquired relevance far beyond their own national confines. See J. Bandopadhyaya, Mao-tse-Tung and Gandhi (New Delhi: Allied, 1973).

3. The Collected Works of Mahatma Gandhi, vol. 48 (New Delhi: Government of India, Publications Division), p. 409.

4. Ibid., vol. 24, pp. 123-124.

5. Ibid., vol. 18, p. 26.

6. Ibid., vol. 29, p. 18.

7. See Pyarelal, New Horizons (Ahmedabad: Navajivan, 1959).

8. Collected Works, vol. 62, p. 29.

9. Quotations in this discussion until note 10 are to be found in Harijan, vol. 6 (New York: Garland, 1973), sequentially, pp. 114, 443, 397, 352, 282, 290, 395; vol. 7, p. 265.

10. Krishanlal Shridharani, The Mahatma and the World (New York: Duel, Sloan and Pearce, 1946), p. 173.

11. D.G. Tendulkar, Mahatma, vol. 5 (New Delhi: Government of India, Publications Division), p. 316.

12. Harijan, vol. 7, p. 265.

13. G.D. Birla, In the Shadow of the Mahatma (Bombay: Orient Longmans, 1955), p. 302.

14. Ibid., pp. 250-252.

15. Quotations herein until note 16 are given in Harijan, sequentially, vol. 7, pp. 275-276, 281; vol. 8, p. 261; vol. 7, p. 285; vol. 8, pp. 172, 185, 214, 212, 222; vol. 7, p. 331.

16. Birla, Shadow of the Mahatma, p. 256.

17. Tendulkar, Mahatma, vol. 6, pp. 114-116.

18. Ibid., p. 116; also, Louis Fischer, The Life of Mahatma Gandhi (Stuttgart: Tauchnitz, 1953), p. 326.

19. Harijan, vol. 9, p. 112.

20. Gandhi's Correspondence with the Government 1942-44 (Ahmedabad: Navajivan, 1957), p. 250.

21. Harijan, vol. 9, pp. 12, 76, 88, 104.

22. Ibid., p. 240.

23. Ibid., p. 73.

24. Ibid., vol. 6, p. 381.

25. Collected Works, vol. 48, pp. 420-421.

26. Krishnanlal Shirdharani, War Without Violence (Bombay: Bhartiya Vidya Bhawan, 1962), pp. 49-50; Harijan, vol. 8, p. 90.

27. Harijan, vol. 6, p. 394.

28. Pyarelal, Mahatma Gandhi, The Last Phase, vol. 2 (Ahmedabad: Navajivan, 1958), p. 808.

29. Harijan, vol. 10, p. 8.

30. Ibid., p. 335.

31. Ibid., p. 174.

32. Ibid., p. 169.

33. Ibid., vol. 9, p. 413.

34. Ibid., p. 365.

35. Ibid., vol. 10, p. 128.

36. Ibid., p. 342.

37. Ibid., vol. 8, p. 245.

38. Ibid.

39. Ibid., vol. 7, p. 285.

6
THE MEASURE OF PEACE

The various commonly held definitions of peace, in Gandhi's opinion, were at once too facile and arbitrary to be taken seriously. Peace as merely the absence or opposite of war, or the English connotation of it as simply the cessation of hostilities formally codified in a treaty, was a rather limited view to take of a much broader, dynamic phenomenon; it was also to touch upon just the negative aspects of the phenomenon.

As one who declared peace to be his religion and proclaimed tirelessly that he was "a man of peace" whose faith in it was his all-consuming passion,[1] Gandhi found the popular literal meaning of the term to be thoroughly inadequate and warped. The peace "you find in stone [or] in the grave" was utterly meaningless, futile; "for the English peace is the peace of the grave. Anything could be better than the living death of the grave."

For Gandhi peace was nothing if it were not an absolutely positive value, unhedged by qualifications, unsullied by compromises, and unencumbered by double entendres. Accordingly, to him peace meant the elimination or destruction of all kinds and forms of tyranny. It was a positive condition conducive to the realization, preservation, and promotion of human dignity and social justice. This, and this vital fount alone, rendered Gandhi's concept of peace moral, fulfilling, enduring, and, of course, dynamic.

Most important, for Gandhi, peace was never an end in itself. It was but a means to a nobler goal: that of a just world order. "The way of peace," he maintained, "is the way of truth. Truthfulness is even more important than peacefulness."[2] The pristine quality of his concept led Gandhi to reject the attitude "peace at any cost, certainly not by placating the aggressor or at the cost of honour."[3] He was revolt-

165

ed at the very idea of seeking material, inevitably temporary, security by appeasing the tyrant. "Peace," insisted Gandhi, "must be just."[4] Profoundly, even mystically, he observed, "I do want that peace which you find embedded in the human breast."[5] And he unfolded, elaborated his ideas on peace over the years through candid comments on the issues of the day.

Thus, at the conclusion of World War I, Gandhi observed: "The war has ended but with little result. The hopes it nourished have remained unfulfilled. The peace, which was expected to be a permanent one has turned out to be one in name. The war which was greater than the Mahabharata has been proved to be but a prelude to a still greater war. . . ." Indeed, he feared--and his apprehensions turned out to be correct--that the Allied victory in 1919 was a peril to world peace: "If the defeat of Germany and the Central Powers ended the German peril, the victory of the Allies has brought into a peril no less deadly to the peace of the world." Further,

> If the terms [of the 1919 peace treaties] which are announced had been described as the terms of war, the description would have been truer to facts.
> But, as was said in ridicule of the late Mr. Gladstone by an equally great man the late Mr. Disraeli, language was invented not to express men's thoughts but to conceal them. That remark is perfectly applicable to the peace terms now announced.

In Gandhi's view, to say that there is peace where one party forces the other to agree to something against its will, crushes it under its brute strength, was a gross violation against truth. Among the other all-too-evident things, the British had sadly failed "to secure just and honourable peace for Turkey." So, where was the occasion for Indians to celebrate peace?

Similarly, a couple of decades later, he felt impelled to describe the Munich Pact as "peace without honour."[6] Europe had been robbed of its honor; the price paid was too heavy. Indeed, England and France had sued for "a peace which was no peace." His scathing comment was that "Europe has sold her soul for the sake of a seven days' earthly existence." The peace the statesmen seemed to prize was nothing more than a makeshift bought either by selling one's honor or punishing the vanquished.

In the same vein, shortly before the conclusion of World War II when the explorations for peace were on, Gandhi wrote: "Peace must be just. In order to be that, it must

neither be punitive nor vindictive. . . . The fruits of peace
must equally be shared."[7] He was quite clear that an armed
peace imposed upon the forcibly disarmed was quite hollow
and temporary.

World War II merely lent confirmation to his views. "The
War," he said, "will end this year or next. It will bring victory to the Allies. The pity of it is that it will be only so
called. . . . That victory will be assuredly a prelude to a
deadlier war. . . ."[8] The same sentiment was reflected in
his "congratulations" to Eleanor Roosevelt: "Your illustrious
husband died in harness and after war had reached a point
where allied victory had become certain. He was spared the
humiliating spectacle of being party to a peace which threat-
ens to be a prelude to war bloodier still, if possible." When
Ralph Coniston, of the Colliers weekly, asked Gandhi why
was he so pessimistic about the possibility of lasting peace
emerging from the defeat of the Axis powers, Gandhi's
prompt response was:

> Violence is bound sooner or later to exhaust
> itself but peace cannot issue out of such ex-
> haustion.
> I am uttering God's Truth when I say that
> unless there is a return to sanity, violent
> people will be swept off the face of the earth.
> . . .
> Those who have their hands dyed deep in
> blood cannot build a non-violent order for the
> world.

Apparently, Gandhi felt quite clearly that war was just
not the means to peace. In fact, his deep-rooted conviction
had made him steadfastly declare that he was an "uncompromising opponent of violent methods even to serve the noblest
of causes."[9] War, for him, thus automatically became a highly dubious--indeed, an illegitimate and morally indefensible--
means for achieving peace. How indefatigably he had forever
proselytized that progress toward a goal must be in direct
proportion to the purity of means! This fundamental philo-
sophic position--strengthened by overwhelming evidence offered by history that war had not succeeded in giving humanity an enduring peace--impelled Gandhi to the simple proposition that peace just could not be attained through force and/
or compulsion.

The logic of his fundamental position, accordingly, led
Gandhi to hold that a peace concluded through war was not
real peace; it was at best a patched-up situation resulting
from mutual exhaustion. Such a temporary affair, and such
methods, could possibly not create a new world order.[10] Indeed, even a "victory attained by violence is tantamount to

defeat, for it is momentary."[11] Only when purer methods
were employed was victory greater and peace more inevitable
and permanent. For Gandhi, only the things attained through
love and peace were imperishable, lasting in contrast to
those gained through the force of arms, or other brutal
methods:

> Hitler and Mussolini on the one hand and Stalin
> on the other are able to show the immediate ef-
> fectiveness of violence. But it may be as transi-
> tory as that of Chenghis' Slaughter.
> But effects of Buddha's non-violence persist
> and are likely to grow with age. And the more
> it is practised, the more effective and inex-
> haustible it becomes and ultimately the whole
> world stands agape and exclaims 'miracle has
> happened.'[12]

To think that peace could be established by means of
war was, thus, to labor under a serious--and disastrous--
delusion, Gandhi maintained. "Soviet Russia believes its violence is a transitional stage to the establishment of an order
without violence," he pointed out, but "the objective is no-
where near fruition." Had violent methods or war the capability of bringing about peace, he asserted, there would not
have been the succession of wars the world has had to suffer.

> I reiterate my conviction that there will be no
> peace for the allies or the world unless they
> shed their belief in the efficacy of war and its
> accompanying terrible deception and fraud and
> are determined to hammer out real peace based
> on freedom and equality of all races and nations.
> Exploitation and domination of one nation
> over another can have no place in a world
> striving to put an end to all war. In such a
> world only the militarily weaker nations will be
> free from the fear of intimidation or exploita-
> tion.[13]

An evil returned by another evil, Gandhi argued, succeeded
merely in multiplying it, rather than mitigating it. It was a
universal law that violence could only be quenched by non-
violence or nonresistance, rather than by superior forms of
violence.

Gandhi had not a shadow of doubt, ever, that the only
way to peace was through nonviolence and nonresistance.
The pity of it was that the true meaning of nonresistance
had ever so often been misunderstood--even distorted. It

never implied, for instance, that a nonviolent being should
bend before the violence of an aggressor. According to Gandhi, in such a situation the true meaning of nonresistance
lay in refusing to submit to the illegitimate demands of the
aggressor, even to the point of death, rather than returning
violence with violence. Further, were an adherent of nonviolence to be confronted by a demand, made with the threat
that violence would follow if the claim were not admitted--say,
a demand like the claim for Pakistan--Gandhi would first like
to ascertain whether the demand had any merit or was worth
striving for. If it were worthy and just, he would support
and propagate it. If, on the other hand, it were not, he
would resist it nonviolently, neutralize it by withholding his
hand. That, according to Gandhi, was the only civilized mode
of conduct in the world. Any other course would ensure a
race for armaments, interspersed possibly with short spells
of tentative peace, the latter being the manifestations of exhaustion during which preparations for violence of a superior
order went on, nonetheless.[14]

ON "PEACE" TREATIES

In the prevailing self-aggrandizing international environment rife with implacable deep-rooted mutual suspicion,
Gandhi could not possibly muster much faith in the peace
pacts of the day, the treaties or agreements pretending to
lead to durable peace.

Accordingly, the Treaty of Versailles, in his opinion,
was "a treaty of revenge against Germany by the victors."[15]
The Germans naturally had a rankling sense of injustice and
insult meted out to them by the great powers. In support of
his contention, Gandhi said, "Only the other day an esteemed
English friend owned to me that Nazi Germany was England's
sin, and that it was the Treaty of Versailles that made Hitler." Thus, even though the immediate responsibility for
World War II might be said to lie with Hitler, it would be
nothing short of crass stupidity, felt Gandhi, to put the
whole blame on the German leader.

Gandhi was emphatic that basically the fault lay in the
whole approach to the issues of war and peace. The Allies
in 1919 made Germany solely guilty of war and showed neither
understanding nor consideration, let alone mercy, in their
judgment of that nation. But did their approach prevent
World War II?

With regard to the treaty with Turkey, Gandhi charged:

> Indeed the terms of the so called peace with
> Turkey, if they are to last, will be a monument
> of human arrogance and man made injustice. To

> attempt to crush the spirit of a brave and gal-
> lant race, because it has lost in the fortunes of
> war, is a triumph not of humanity but a demon-
> stration of inhumanity.[16]

From his fundamental differences with the prevailing approach flowed Gandhi's serious doubts with regard to the attempts for peace by the European statesmen during the interwar years. His visit to Europe in 1931 revealed to him "a
greater longing for peace on the part of the people" there;
underneath the great material progress, however, simmered
intense dissatisfaction and unrest among the people. It was,
in his opinion, a healthy sign. But, at the same time, "appearances . . . show that the world [is] preparing for another war, [though] one must hope that it may be possible
to avoid it."[17] Accordingly, Gandhi was least impressed by
the peace moves leading to the Kellogg-Briand Pact of August
1928. The pact proclaimed that the signatories would renounce
war as an instrument of national policy. But Gandhi had no
illusions about it for the simple reason that the subscribers
to the pact were, by and large, partners in exploiting the
peoples of Asia and Africa. The pact was thus merely an
arrangement "to carry on the joint exploitation peacefully;"
at best, it was an attempt to preserve status quo and nothing
else. Similarly, the Munich Pact, Gandhi believed, merely
postponed the war: "the danger of war has been averted for
the time being." England and France had failed to secure
anything, least of all peace. The pact, he asserted, marked
Hitler's personal triumph of violence.[18]

Gandhi also thoroughly disapproved of secret treaties
signed during war, for such treaties, concluded as they
were in the darkness of intrigue and pressures of expediency, could not ensure peace. They were convenient arrange-
ments for achieving a particular objective, but once the war
ended they became meaningless. In fact, even the signatories
to such treaties considered "their own written word as so
much waste paper."[19] He viewed the German-Soviet Pact of
August 23, 1939, as an unnatural combination of the two
countries in a seemingly happy, but quite unintended, fusion
the ultimate shape of which nobody knew.[20] "Since the 1919
Treaty of Peace was signed . . . peace seems almost as far
as ever from resuming its sway over mankind."[21] Because
it was nothing more than a treaty of revenge, "Messrs Hitler
and Company have reduced to a science the unscientific violence their predecessors had developed for exploiting the so-
called backward races for their own material gain."[22] Absolutely nothing could persuade Gandhi to the contrary.

Nor did he believe in the ability of international bodies
to ensure peace. There was no sense, or use, in making appeals to these organizations, for they were entirely unable to

do justice. Referring to the Syrian problem after World War I, Gandhi held:

> And what is the League of Nations? Is it not in reality merely England and France? Do other powers count? Is it any use appealing to France which is denying her motto of Fraternity, Equality and Justice? She has denied justice to Germany, there is little fraternity between her and the rights and the doctrine of equality she is trampling under foot in Syria.[23]

Since it was pointless, in the opinion of Gandhi, to invoke the protection or help of the League of Nations, he advised the Abyssinians not to bother to "appeal to the League or any other power for armed intervention" against the Italian aggression.[24] Giving expression to his dim view of the League, he said:

> The League is expected to perform wonders. It is expected to replace war and by its own power arbitrate between nations who might have differences between themselves.
>
> But it has always seemed to me that the League lacks the necessary sanction. It depends as it has to, largely if not exclusively, on the judgment of the nations concerned.
>
> I venture to suggest to you that the means we have advocated in India supply the necessary sanction not only to a body like the League but to any organisation for this great cause of the world.[25]

Gandhi had, in fact, had serious doubts about the sincerity of the League of Nations almost from its very inception --and with good reason. As early as February 25, 1919, his letter to Deenabandhu C. F. Andrews drew attention to the hypocrisy of the founders in defining the intent of the organization:

> Have you noticed an unconscious betrayal of the true nature of modern civilisation in Mr. Wilson's speech explaining the League of Nation's Covenant? You will remember his saying that if the moral pressure to be exerted against a recalcitrant party failed, the members of the League would not hesitate to use the last remedy, viz., brute force.[26]

Paralysis into inaction and sheer lack of will on the part of the League of Nations with regard to the Japanese aggression in Manchuria in 1931 had further led Gandhi to hold a rather poor opinion of the League's capabilities: the organization was hopelessly weak and utterly futile so far as the maintenance of peace among nations was concerned. The collective security system the League had devised merely facilitated pursuit of national interests, especially of the dominant powers. Eliminating war or ensuring peace, consequently, became quite marginal--if that.

International conferences concerning themselves with some aspect of a peaceful world order, similarly, scarcely aroused Gandhi's enthusiasm or faith. The communist-inspired League Against Imperialism organized an Oppressed Nations Conference at Brussels in 1928. Jawaharlal Nehru attended that conference and communicated his impressions of it to the mahatma, whose response was, "I read most carefully your public report as also your private confidential report about the doings of the Oppressed Nations Conference. I myself do not expect much from this League [Against Imperialism]."[27] Gandhi felt that the capability and activity of such a league was virtually dependent upon the goodwill of the powers, which were very much partners in the exploitation of the oppressed peoples. The European nations that joined the League Against Imperialism would find themselves unable to resist pressures, persuasions, and temptations of the exploiters for the obvious reason that the European nations would just not be able to accommodate themselves to what in their opinion amounted to an injury to their self-interest. The oppressed nations, as a result, would look to external help instead of generating their own strength for achieving their objectives of freedom and peace.

Gandhi was, similarly, highly skeptical about the great powers, deliberating at San Francisco to bring forth an agency or method capable of guaranteeing peace: "I am positive," he asserted, "If they are so arrogant as to think that they can have lasting peace while the exploitation of the coloured and the socalled backward races goes on, they are living in a fool's paradise." He was not even sure that the Allies will stand united for long:

> The quarrel with Russia has already started. It is only a question when the other two--England and America--will start quarrelling with each other. May be, pure self interest will dictate a wiser course and those who will be meeting at San Francisco will say: let us not fall out over a fallen carcass.[28]

The security system worked out in San Francisco, he was sure, would not guarantee peace, as "the conference will have much to do with the world to be after the so-called end of the War. I very much fear that behind the structure of the world security sought to be raised lurk mistrust and fear which breed war."29

In short, world peace, Gandhi believed, could never be established through conferences: "peace is being broken as we all see even while conferences are being held."30 The world organizations--League as well as the United Nations-- were, essentially, the products of war and not positive manifestations of the natural urge of nations for peace. These institutions--basically the result of lust, anger, fear, mistrust, selfishness, and compromise--thus could not possibly preserve or promote peace. They could not rise beyond superficialities and limited, mechanical devices like "collective security systems" for the motives of their originators were negative and base: "The United Nations set out to fight Hitler with his weapons and ended by out-Hitlering [Hitler]."

All this only buttressed Gandhi's own conviction with regard to the means suited to ensuring peace: "War will never be terminated by any agency until men and nations become more spiritual and adopt the principle of brotherhood and concord rather than antagonism, competition and brute force."31 According to Gandhi, "only an organisation based predominantly on truth and non-violence" could promote and preserve peace.32 Once the Western nations had sincerely recognized the power of spirituality, Gandhi was certain, they will be free from war, the crimes of violence, and the things that go with these evils.

However, with the world, and the states in it, being what they were, how would Gandhi resolve the differences and disputes among the peoples of the world? He recommends arbitration; war was certainly not the way to tackle interstate disputes. Thus Gandhi desired Hitler, for instance, to accept arbitration with regard to the latter's claim to Danzig and the Polish Corridor:

> It is highly probable that his right to incorporate Germany is beyond question, if Danzig Germans desire to give up their independent status.
> It may well be that this claim be a just claim. My complaint is that he will not let the claim be examined by an independent tribunal33 [emphasis added].

Gandhi seemed willing to concede that the League of Nations might not be a proper tribunal in this particular case. To spare the world the savagery of war, Gandhi greatly wished that Hitler "would respond to the appeal of the President of the United States and allow his claim to be investigated by arbitrators in whose choice he will have as effective voice as the disputants"!

Arbitration, again, was the method he suggested for settling the issue of war debt between Britain and free India. He seemed inclined to submit even the question of Pakistan to arbitration if no other solution was possible.[34] Similarly, in the dispute between India and Pakistan over Junagarh, Gandhi advocated that the only honorable way out was the "ancient method of arbitration in the usual manner." There were enough men and women in India who could shoulder the burden. If, however, arbitration by Indians were unacceptable, Gandhi had no objection to any impartial person from any part of the world.[35]

However, just as he would not even consider arbitration by the League of Nations or any other agency on the question of India's independence,[36] Gandhi did not favor this procedure in the Indo-Pakistan dispute over Kashmir.[37] He wanted the two countries to sort out the Kashmir problem amicably through their mutual efforts, but not by means of war, for "a war would bring both [nations] under the sway of a third power and nothing could be worse." He pleaded for amity and goodwill so that India's representation to the Security Council on Kashmir could be withdrawn with dignity. Gandhi was inclined to think that the United Nations itself will welcome such a step and would favor a mutual settlement of the problem, for "to harbour internal hatred might be even worse than war." Vincent Sheean has recorded that Gandhi spoke to him of the failure of the United Nations in the matter of Kashmir.[38]

PACIFIST CONSTRAINTS ON PEACE

Gandhi's views on peace sharpened in his debate with the Western pacifists, especially in the interwar years. In contrast to the limited antiwar posture of the pacifists, Gandhi went to the very root of the problem when he focused entirely on the causes of war, than the war itself. He held the inhuman race for the exploitation of the weaker peoples of the world to be primarily responsible for war. Accordingly, he told the pacifists, "all activity for stopping war must prove fruitless so long as the causes of war are not understood and radically dealt with."[39]

Gandhi was of the firm conviction that unless the very spirit of the nations changed, mere slogans or campaigns for the eradication of war were futile, as the following statement indicates.

> Will America, England and the other great na-
> tions of the West continue to exploit uncivilized
> races and hope to attain peace that the whole
> world is pining for? Or will Americans continue
> to prey upon one another, have commercial
> rivalries and yet expect to dictate peace to the
> world?
>
> Not till the spirit is changed can the form be
> altered. The form is merely an expression of
> the spirit within. We may succeed in seemingly
> altering the form but the alteration will be a mere
> make believe if the spirit within remains unalter-
> able. A whited sepulchre still conceals beneath it
> the rotting flesh and bone.[40]

Herein lay, according to Gandhi, the basic reason of the
utter ineffectiveness of the pacifist movements in the world.
When the British pacifists, for instance, "speak of pacifism,
they do so with the mental reservation that when pacifism
fails, arms might be used. With them not non-violence but
arms are the ultimate sanction, as was the case with Woodrow
Wilson's Fourteen Points."[41] But war will be eradicated only
if personal greed were killed. And in that respect, believed
Gandhi, the war resisters had nothing whatever to say. The
prospect of armed conflict between nations is horrifying, of
course, he admitted,

> but the economic war is no better than an
> armed conflict. This is like a surgical opera-
> tion. An economic war is prolonged torture.
> And its ranges are no less terrible than those
> depicted in the literature on war. . . . We
> think nothing of the other because we are
> used to its deadly effects.
>
> The movement against war is second. I
> pray for its success but I cannot help the
> gnawing fear that the movement will fail, if it
> does not touch the root of all evil men's greed.[42]

It is not surprising, then, that when Gandhi was invited to
sign the "response" to President F. D. Roosevelt's advocacy
of the Moral Rearmament Movement, he refused, for he could
not endorse what he considered to be a falsehood: "The
whole appeal is so unreal. . . . Let them shed their exploitation and their immoral gains first," he demanded.[43]

Gandhi did not, however, entirely ignore or underesti-
mate the valiant endeavors of the pacifists "to kill war spirit." In fact, he was quite impressed by the zeal of certain
religious and other groups working for peace. In various
issues of Young India, he published Kirby Page's essay on

the causes and remedies of war. Commenting on the suggested remedies, Gandhi observed, "The reader will find the writer weak in his statement of remedies, not because he is weak in his belief, but because it is new ground for everybody. Nobody wants war. But how can an age-long institution be easily destroyed?" And exercising an editor's privilege, he changed Page's caption "What Shall the Churches Do About War?" to "How Can Religion Help?" which he considered to be a more suitable title of one of the remedies suggested by Page. Gandhi believed that all the religious-minded could definitely help in finding the remedy for war; in fact, they actually were doing so, in a way, as evidenced by the activities of the Fellowship of Reconciliation.

By the same token, the anticonscription movement enjoyed Gandhi's support and blessings, notwithstanding his skepticism of such activities: "the reformers will have to put up an immense struggle to secure State action in the desired direction. Each is afraid and distrustful of his neighbour." But he agreed that war could be avoided if people and military forces did not extend their cooperation, and such "non-co-operation in case of war is every where possible and it is by this that universal peace will be obtained." Besides his understandable preference for the method of noncooperation, Gandhi persisted in holding that the exploitation of nations constituted a crucial cause of war. "There can be no living harmony between races and nations," he said in his message to a French pacifist, "unless the main cause is removed, namely, exploitation of the weak by the strong. We must revise the interpretation of the so-called doctrine of 'Survival of the Fittest.'"[44] As a positive, and practicable, first step in that direction, he emphasized the necessity of removing the existing disparities among nations; equality between man and man will ensure peace, he maintained: "The moment we have restored real living equality between man and man, and the whole creation, when that day comes we shall have peace on earth and goodwill to men."

Gandhi found the Marxists' panacea--abolish the institution of private property and war will automatically be eliminated--too facile and naive. "To banish war," he said, "we have to do more. We have to eradicate possessiveness and greed and lust and egotism from our own hearts. We have to carry war within ourselves to banish it from society."[45] He did not think that the Soviet Union was entirely free of expolitation and, thus, did not want "Russia as a model to be emulated." One reason, among others, was that the Soviets relied on force and violence;[46] "let nobody be misled by Russian parallel," he cautioned.[47]

Racism, in Gandhi's opinion, was yet another factor that breeds war; South Africa's policy in this regard "holds the seeds of a world war." He declared, "Those who agree that racial inequality must be removed and yet do nothing to fight the evil are impotent. I cannot have anything to say to such people. After all, the underdogs will have to earn their own salvation."[48] And he was confident that India had a decisive role to play in this regard--"to play an effective part in straightening up affairs," as he said. He hoped that the United Nations might solve the problem of apartheid or racialism, but "If the UNO fails to deal justly with the South African . . . dispute, it will lose its prestige. I have no doubt that the UNO can prosper only if it is just." And if justice were not done, "One day the black races will rise like avenging Attila against their white oppressors, unless some one presents to them the weapon of Satyagraha."[49]

Gandhi did not experience much of the cold war; he lived only during the first chilling drafts of it. He anticipated it with considerable perspicacity, of course, because the faith he had in his own analysis of war remained firm; nothing had happened to disprove, discard, or revise the analysis. The aftermath of a war, according to him, could be nothing but

> deeper hatred and counter-hatred, and vengeance let loose. . . . The recent war, whose embers have yet hardly died, loudly proclaims the bankruptcy of this use of hatred. And it remains to be seen whether the so-called victors have really won or whether they have not depressed themselves in seeking and trying to depress their enemies.[50]

Gandhi had long subscribed to this view. Even as the war was on, he had advocated that retaliation should at all costs be avoided as an effective method of terminating the war. "War could take place only if there are two parties to fight one another,"[51] he said, implying that the spirit of retaliation automatically causes an adversary to materialize quickly. Were retaliation to be resorted to, parties prepared for war would forever be there.

He found abundant support in history for his viewpoint that evil feeds on resistance; history also furnishes many instances where people have tamed the fiercest with their all-embracing love. Of course, he fully appreciated that "such non-resistance requires far greater courage than that of a soldier who returns two blows against one," and not many were capable of it. So in cases where people have anger instead of love in them for the evil-doer, he argued, it is better for them to fight clean, rather than, in a cowardly manner, to sit still for fear of dying. In Gandhi's vocabulary, cowardice and brotherhood were absolutely contradictory terms.

Gandhi found it to be nothing short of blatant hypocrisy to proclaim faith in brotherhood while not being prepared to

love even our enemies. According to him those who had imbibed the spirit of brotherhood should be able to say honestly that they have no enemies. Gandhi was acutely aware that
there was no acceptance of this in the world; in fact, his
plea for nonretaliation might well be rebuffed. But he was
absolutely sure that in the absence of nonretaliation there
could be no peace in the world, and, under the circumstances,
"to take the name of brotherhood [was] a blasphemy." Yet to
be able to refrain from retaliation was not all that impossible
as one might imagine. The reason why doubts rise in our
minds, stated Gandhi, is that thus far we have not realized
our humanity or dignity, nor have we quite shed some of the
traits inherent in the Darwinian concept of the descent of
man. Citing the example of Quakers and Doukhobors, who
lead the life of nonretaliation, Gandhi maintained that what
"is true of individual can be true of nations or of groups of
individuals," not at once perhaps but certainly in a gradual,
evolutionary process. He had a fundamental quarrel with those
who believed that people cannot reach the stage where they
could do without retaliation. Gandhi was confident that we
certainly can realize our full destiny and dignity only if we
educate and train ourselves to be able to refrain from retaliation.

The obvious, compelling implication of nonretaliation is
nonviolence. And that, according to Gandhi, was perfectly
valid--indeed, eminently suited--for international conduct:

> I do justify entire non-violence and consider it
> possible in relations between man and man and
> nations and nations, but it is not a resignation
> from all real fighting against wickedness. On
> the contrary the non-violence of my conception
> is a more active and more real fighting against
> wickedness than retaliation whose very nature
> is to increase wickedness.[52]

Gandhi was, evidently, contemplating a mental and moral opposition to immoralities prevailing in the world. Because
"peace may arise out of strife, all strife is not [automatically]
anti-pacific. To stand with folded hand is not to achieve reform." He believed that, instead of returning violence with
violence, its neutralization by withholding one's hand and
simultaneously refusing to submit to the aggressor's forceful
demand was the only civilized way of getting on in the world.
"There is no peace," he said, "for individuals or for nations
without practising truth and non-violence to the uttermost
extent possible for man. The policy of retaliation has never
succeeded." Gandhi dismissed the argument that nonviolence
takes a long time to be effective due to the lack of will and
sincerity of its adherents. Nor did he agree that it was the

opening of the second front by Hitler in June 1941 that saved the Soviet Union. According to Gandhi, nonviolence in that situation would have been as effective, if not more: "It is not permitted me to think or else I would be denying my faith which today burns brighter than ever."[53] The faith alluded to here was none other than his conviction that nonviolence was the only way for humanity.

THE PATH TO PEACE: AN INDEPENDENT INDIA

Gandhi's terse prescription for peace through operationali- zation of his ideas was: free India. That would provide the medium through which peace could be attained and a new world order created: there could not be any peace in the world as long as India was denied independence and kept in shackles. The motto he had set for a free India was: "If we're to be saved and are to make a substantial contribution to the World's progress, ours must emphatically and predomi- nantly be the way of peace."[54] He hoped that his country would attain freedom through the spiritual force of satyagraha. The next step would be to spread the tenet and technique of this highly efficacious, constructive, and peaceful force: "Civilisation can be saved only through Satyagraha" now and forever, Gandhi proclaimed.

It may be stressed once again that satyagraha is a unique- ly Gandhian concept and technique. Qualitatively much more positive than mere noncooperation, civil disobedience, or non- violent resistance, satyagraha calls for and strengthens high moral fiber in the individual. As its literal meaning suggests, it connotes the aagraha (compulsion) of satya (truth, reality). It promotes and strengthens the satya it springs from; in- stead of destroying it builds; out of dissonance and chaos it creates resolve and enrichment. It is the ultimate instrument for constructiveness, peace, and dynamism.[55] Gandhi had arrived at the value of satyagraha from personal experience, and this is how he elaborated his position.

Gandhi struggled so that India could attain its indepen- dence, not at the expense of other nations, nor in a manner that led to the extinction of other nations, but through its own efforts and strength and by totally nonviolent means. He was confident of India winning its freedom through a gentlemanly understanding with Britain, with dignity acquired by the practice of nonviolence. India's nonviolence would con- vert an imperialistic Britain maneuvering for world supremacy to a Britain humbly trying to serve the common goal of human- ity. Once free, India would no longer be driven helplessly into Britain's wars of exploitation. Instead, India would be a powerful nation seeking to restrain all the violent forces of the world. "Through realisation of freedom of India," said

Gandhi, "I hope to realise and carry on the mission of the brotherhood of man." India must rise free "so that the whole world may benefit. I do not want India to rise on the ruins of other nations," he said. Fully appreciating that no country could ever remain in isolation for any length of time in the world today, he envisaged a role of vigor and activity for a free India:

> The present plan for securing Swaraj is to attain not a position of isolation but one of full self-realisation and self-expression for the benefit of all.
> The present position of bondage and helplessness hurts not only India, not only England, but the whole world.[56]

This explains his appeal "To Every Briton" during World War II, urging British withdrawal "at this very hour" from every Asiatic and African possession including India so as to ensure destruction of fascism and nazism and the saving of the world.[57] Talking to an American in 1940, Gandhi pointedly said: "Your wars will never ensure safety for democracy. India's experiment can and will, if the people come up to the mark, or to put it another way, if God gives me the necessary wisdom and the strength to bring the experiment to fruition."[58]

By making the freedom of India the key to an enduring world peace, Gandhi was seeking to change the outlook of the whole world toward war and peace, particularly when he insisted unequivocally that India's independence must come through nonviolence.

"India's freedom," he repeatedly declared in the columns of Young India, "must revolutionize the world's outlook upon peace and war. Her impotence affects the whole of mankind."[59] Subject and silenced India was a menace to the world, not merely because it had no voice in the affairs of the world, but also because a helpless India excited "the jealousy and greed of other countries which must live by exploiting her." The triumph of nonviolent means in the attainment of India's freedom, Gandhi was confident, would change the values of the world: "most of the paraphernalia of war would be found to be useless." India's nonviolent victory, moreover, would inspire and influence the rest of the exploited world; the nation would naturally become a mode! to be emulated by the struggling subject peoples the world over.

Thus India's freedom became Gandhi's way to world peace. With irresistible logic he declared: "I who belong to a subject nation, did not know how I could work for peace except by working for freedom, and if India could be helped to win freedom through peaceful means, it would be a very good combination for peace."[60] And so, while he found the "Four Freedoms" of the Atlantic Charter, of August 1940, hollow and worthy of his cold but dignified silence, he spurned invitations from the United States and China to deliver his message of peace. He believed he could best serve peace by carrying on his mission in India. Similarly, on being invited by some U.S. citizens to become the world leader for the realization of a worldwide commonwealth of spirit, in which humanity's age-old dreams of spiritual progress would characterize the daily life of average people everywhere, Gandhi decided he could best contribute in this regard by "restricting his experiment to a fragment [India]."[61]

His response to the world propaganda for eradicating war was, again, typical:

> I doubt the utility of an organisation for the education of public opinion by various kinds of propaganda "for eradicating war". That propaganda has really no meaning in our country at the present moment.
>
> I have repeated letters from America from the Fellowship of Reconciliation. I am in correspondence with them still but I have not joined them as it seems to me to be a mockery for me to join.
>
> May a mouse with any propriety join the organisation run by cats for the purpose of stopping war on mice?
>
> It is therefore enough for us to realise our status and "pray in secret" for peace on earth.

And with his customary candor, he pointed out further:

> I cannot save Europe and America. If I go there, I shall be like a stranger. Probably I shall be lionised but that is all. I shall not be able to present to them the Science of peace in a language they can understand. But they will understand if I can make good my non-violence in India.[62]

We may point here to what we think was a sterling trait in Mohandas Karamchand Gandhi, namely, irrepressible honesty and sincerity--at worst, modesty and humility--as the foundations of his thinking and conduct; he instinctively recoiled from flamboyance or ostentation. Sounded to be recommended for the Nobel Prize for Peace, Gandhi promptly told his friends to desist from any such step for, to him, his endeavors in the cause of world peace were in themselves their own reward. Any recognition or respect for nonviolence in Europe or elsewhere would, of course, be very welcome to him. But the value of such a recognition would instantly be lost if instead of a spontaneous gesture the award of the prize had to depend upon extraneous recommendations by friends and well-wishers.[63]

With regard to the value and actual practice of nonviolence in non-Indian environments, Gandhi was not aware of the peculiarities in those places, which might well pose some problems of their own. "Friends have told me," he observed,

> there were special difficulties in Europe to adopt non-violent means. Europe consists of martial races unlike India. Here all know how to wield arms. All the war population has at one time or another wielded arms.
>
> It is difficult for you to understand the efficacy and beauty of non-retaliation. Why not punish the wrong-doer--and in an exemplary manner--that is what is asked everywhere here. Thus non-violence is quite foreign to Europe. For people belonging to such a country (sic) it is difficult to strike out a new patch. Your economic life is so constructed that it is not possible, generally speaking, for an ordinary man to get out of the ordinary, unless he faces poverty.
>
> And the fourth difficulty is that in catholic Europe the iron discipline allows very little free play to the intellectual.
>
> These are the four difficulties we have not to face.[64]

But he never considered these difficulties to be insurmountable. Knowing the world to be what it was, he was too much of a hard-headed realist to have expected an easy or smooth acceptance or practice of nonviolence, in any case. That is why he wanted the Indian experiment with nonviolence to be a resounding success to inspire, and serve as a model to, the world. And he bent all his energies to make the experiment a success in India.

Gandhi could not even think that India could ever be turned into a military camp like Germany, for instance; it would take ages to do that, he said. To teach the people in an organized manner to remain passive, that is, pacific under suffering, was relatively easier in India. He was absolutely clear about the kind of India he wanted to mold:

> If India takes up the doctrine of the sword,
> she . . . will cease to be the pride of my
> heart. . . .
> I believe absolutely that she has a mission
> for the world. She is not to copy Europe blind-
> ly. India's acceptance of the doctrine of the
> sword will be the hour of my trial. I hope I
> shall not be found wanting.[65]

Another statement by Gandhi reaffirms this stand: "If India becomes free through non-violent means it won't enter upon war. But if she does, God will give me the strength to fight India singlehanded." This, to say the least, was the measure of his faith that a free India would be a haven of nonviolence and a beacon for peace in the world, not a menace but a blessing to the rest of humanity. Such an India would be obliged to perfect the instrument and content of nonviolence for resolving conflicts; otherwise its freedom would be worthless.[66] Almost poetically he said, "If India acts on the square when her turn comes, it will not dictate terms at Peace conference, but peace and freedom will descend upon it, not as a terrifying torrent, but as 'gentle rain from heaven.'"[67]

Gandhi was acutely and painfully aware that India had yet not attained the heights of nonviolence he had set for it. "Not until the Congress, or a similar group of people represents the non-violence of the strong, will the world catch the infection." The heinous dance of naked violence and mass massacres of 1947 in his country greatly disheartened him:

> India is becoming the laughing stock of the
> world. . . . The world asks, where is your
> non-violence with which you have won your
> independence? I have to hang down my head
> in shame.
> Will a free India present to the world a
> lesson of peace or of hatred and violence of
> which the world is already sick unto death?[68]

He had been spared the excruciating agony almost up to the last months of his life, of course, but on quite a few occasions even earlier also he had been chagrined over the setbacks that nonviolence suffered in India. These lapses, however, convinced him even more about the relevance of nonviolence while making him still more resolute in inculcating it among his countrymen.

Gandhi knew perfectly well that after World War II Britain was not the only power the weaker peoples had to reckon with: the other Big Three had emerged. The weak would just not be able to engage these powers with their advanced weapons and techniques; "after all you cannot go beyond the atom bomb." So unless a new method of fighting imperialism were developed and adopted, there seemed no hope for the oppressed peoples the world over. In this search, and struggle, Gandhi assigned India a leading role, for a nonviolent India could lead the world toward peace. If, on the other hand, India unfortunately chose the path of militarization, it would become a fifth-rate power:

> What place will India have in the comity of nations? Will she be satisfied with being a fifth rate power? . . . India will have long to wait before she can become a first class military power. And for that she will have to go under tutelage of some western power.[69]

Gandhi questioned whether India would be content to remain a fifth-rate power without having a message for the world or if it would "by further refining and continuing her non-violent policy, prove herself worthy of being the first nation in the world using her hard won freedom for the delivery of the earth from the burden [of violence] which is crushing her in spite of the so-called victory [of the Allies]."[70]

Gandhi's confident answer was that nonviolence alone can solve all our ills and be triumphant in providing a solution to world problems. His supreme confidence in the brilliance of his technique made Gandhi hold that even

> an India reduced in size but purged in spirit may still be the nursery of the non-violence of the brave and take up the moral leadership of the world, bringing a message of hope and deliverance to the oppressed and exploited races.
>
> But an unwieldy, soulless India will merely be an imitation, and a third rate imitation at that, of the western military states, utterly powerless to stand up against their onslaught.[71]

A more realistic evaluation of free India's power and position in the world could hardly be formulated. Even in terms of realpolitik, therefore, nonviolence was the obvious way indicated for an independent India.

Gandhi left the world in little doubt that he had chosen the path of nonviolence for winning the freedom of India not because India was powerless to draw the sword, but because with him "to draw the sword was a sin."[72] Perhaps more than that, the way India chose did not mean cowardice or helplessness. On the contrary,

> It means the spirit of manliness in its perfec-
> tion . . . non-violence--peace does not mean
> submission to others, does not mean weakness.
> He alone can forgive who is brave. When the
> hour comes for bidding goodbye to the Govern-
> ment, India will not be lusterless as she is to-
> day but will be radiating light in all directions.

Moreover, in contrast to the method of violence or war, nonviolent freedom struggles involve no loss of money or people,
at any rate far less--and there is no residue of bitterness.[73]
"Victory," maintained Gandhi, "will come only with non-violence, and . . . whatever good results violent methods may
seem to promise, ultimately they bring defeat. The ruddy
cheeks of a tuberculous patient are not a sign of health, but
a portent of imminent death."

Genuine and enduring victory in the struggle for freedom, or struggle of any kind for that matter, materializes
only by means of nonviolence. India valued this throughout
its history. The Indian ethos and experience thus encouraged
Gandhi to forecast confidently that the method of attaining
its independence would make India work for a "peaceful atmosphere in this troubled globe of ours." And, since India
has "never waged war, unlike other nations she does not
have to develop the will for peace; she has enough [of such
will]."[74]

Independent India's role in international affairs has amply
justified Gandhi's confidence in this regard. Pioneering the
philosophy, posture, and policy of nonalignment by India
broke new ground in international politics. Here emerged a
mode of behavior and a code of conduct for a nation that, instead of exacerbating the game of power politics and brink-
manship, succeeded in lessening tensions and promoting dialogue among angry opponents. Initially ridiculed and rejected
by the powers that mattered on the crucial decision of war
and peace--in fact, on the question of the very survival of
humanity and the planet earth--the appeal of nonalignment
became irresistible enough to make it a significant force in
the world. Indeed, the very morality of this phenomenon, we
suggest, ushered in, inter alia, detente and the subsequent
transformation of the international scene. It would really be
very difficult for anybody to argue that nonalignment has
not been a factor and force for peace, for undoubtedly it
has been, notwithstanding certain setbacks (for instance, the
Chinese military action against India in 1962). Had nonalignment not taken shape, and pursued with the vigor it was
during the 1950s, the world might well have ended in the
smoke of a final world war.

With equal perception, Gandhi also felt that "India's freedom through non-violence will mean a great deal to the whole
world, certainly to all the eastern [Afro-Asian] nations."[75]
India's freedom, it is now almost universally recognized,
touched off the rapid dissolution of empires in Asia and Africa. Freedom came to Burma, Ceylon (Sri Lanka), and many
other countries of Asia and Africa not entirely through armed
struggle or violence, though perhaps not strictly through
Gandhian nonviolence either. At the same time it cannot be
disputed that the means and the manner of transfer of power
did indeed set the tone and the style of freeing the countries
from the yoke of colonialism.

This may not be the place to examine why other nations
failed to adopt nonviolence of Gandhian variety as a means
toward winning independence. But our short response to this
fact, in keeping with Gandhi's own feelings in this regard,
would be that these nations and their leaders did not understand nor attempt to adopt and extend the use of satyagraha
for a variety of reasons. Dramatic changes might ensue if
these peoples would make the effort to comprehend and then
develop the roots of satyagraha in their own cultures to res-
onate with the Indian experience.

Discerning souls in Europe, incidentally, fully appreciated Gandhi's approach. Albert Einstein, for instance, wrote
to Gandhi on September 27, 1931.

> You have shown by all that you have done that
> we can achieve the ideal even without resorting
> to violence. We can conquer those votaries of
> violence by the non-violent method. Your exam-
> ple will inspire and help humanity to put an end
> to a conflict based on violence with international
> help and co-operation guaranteeing peace to the
> world.[76]

Another strand in Gandhi's plea that India attain its freedom nonviolently was that, due to its large size, India could
not possibly adopt the doctrine of force after independence.
"Then India becomes one of the exploiters of the weaker
races of the world. Just imagine, what it must mean to the
world." India must always abjure and resist exploitation and
force, and become invincible by mastering the art and science of nonviolence. India freed by nonviolent means, and
active as a force for peace--while being at peace with the
rest of the world--would not have to bother about aggression
by foreign powers, for it will have no enemies. Even if it
were invaded "without any offence on our part, we must
trust to our capacity for suffering to be able to defend ourselves against all aggressors."[77] India would, hopefully,
have the smallest army imaginable. It is certain that "no one

will dare cast covetous eyes on India" if it is united as a nation, is ever fearless and self-reliant, and possibly is also self-sufficient in its basic requirements.

Gandhi could, of course, not know the exact shape of the policy the national government of India would follow. Maybe it would incline toward militarization. But his earnest advice would be for a full-fledged adoption of nonviolence to the utmost extent. If his advice were accepted by the national government, "all the effort to show the efficacy of non-violence as a political force will . . . not have gone in vain."[78] In that event, "India will become the unquestioned leader of the whole world" too.

Since the freedom of India won through nonviolent means would, presumably, be preserved by the same means, a positive contribution to a durable peace in the world would have been made: "People who have been able to gain liberty through these means are able to retain it through these means, then. . . . India will have made the largest contribution possible to the war against war."[79]

Insofar as India's liberation would touch off the avalanche of decolonization, imperialism—which, as the obnoxious ideology and manifestation of exploitation, was the single largest obstacle to permanent peace—would automatically disintegrate: "India's coming into her own will mean every nation doing likewise." To repeat Gandhi, thus, "an indispensable preliminary to [world] peace is the complete freedom of India from all foreign control."[80] So long as India, and countries like it, were in the exploitative, parasitic meshes of the Allies, there could not be any peace in the world.

All through these pages we have repeatedly drawn attention to Gandhi's deep-rooted, total, and unswerving commitment to and faith in the virtues of nonviolence and how intensely he felt it to be the supreme method: "it benefits both the resistor and the wrong-doer whereas the war harms both the victor and the vanquished."[81] Another citation, substantiating his insistence that India win its freedom by this method, may be offered:

> The way of peace insures internal growth and stability. We reject it because we fancy that it involves submission to the will of the ruler who has imposed himself upon us. But the moment we realise that the imposition is only socalled and that, through our willingness to suffer loss of life or property, we are party to the imposition, all we need do is to change that negative attitude of passive endorsement.
>
> The suffering to be undergone by the change will be nothing compared to the physical suffering and the moral loss we must incur in trying

> the way of war. And the suffering of war harms
> both the parties. The sufferings in following
> the way of peace must benefit both. They will
> be like the pleasurable travail of a new birth.

In view of this, we simply cannot expect Gandhi to favor methods of warfare for liberty both of a people and within a people. He bluntly declared that he would decline to take part in a violent war of independence. He was least inclined toward armed risings or violent revolutions, for, in his opinion, they were a remedy worse than the disease sought to be cured; they were a token of the spirit of revenge and impatience and anger; apart from causing untold misery, they left an aftermath of bitterness and hatred, indemnities and reparations, victimization and reprisals, hanging of traitors and war criminals, and so on. The emergence of Bangladesh in the recent past, for instance, amply illustrates this.

Indian independence, on the other hand, left not a trace of these abominations in its wake, simply because the mode of its attainment was essentially nonviolent. Moreover, India after its independence could become an important, equal, and effective partner working for peace in the world because of its nonviolence. As early as 1920, Gandhi had proclaimed:

> I hope also to show to India and the Empire
> at large that given a certain amount of capacity
> for self-sacrifice, justice can be secured by
> the peacefullest and cleanest means without
> sowing or increasing bitterness between the
> English and Indians. . . . [These qualities of
> self-sacrifice and justice] alone are immune
> from lasting bitterness. They are untainted
> with hatred, expedience or untruth.

Elsewhere, Gandhi declared, "Let the philosophy I represent be tested on its own merit. I hold that the world is sick of armed rebellion."[82] Indian independence vindicated Gandhi completely and confirmed him in the faith that the only way to permanent freedom was not through bloodshed but through nonviolence:

> History shows that when a people have been
> subjugated and desire to get rid of the subjec-
> tion they have rebelled and resorted to use
> arms.
>   In India, on the other hand, we have re-
> sorted to means that are scrupulously non-vio-
> lent and peaceful and strangers have testified
> and I am here to give my testimony that in a

great measure we seem to have succeeded in
attaining our goal.

I know that it is still an experiment in the
making. I cannot claim absolute success as yet
but venture to suggest . . . that experience
has gone so far that it is worthwhile to study
the experience.

I further suggest that if that experience
becomes a full success, India will have made a
contribution towards world peace for which the
world is thirsting.

Gandhi would not endorse the doctrine of the sword, or
violence of any kind, for the defense of even a proven right:
"Proved right should be capable of being vindicated by right
means as against the rude, i.e., sanguinary, means. Man
may and should shed his own blood for establishing what he
considers to be his right."[83] He cited the plight of U.S. negroes as an example: slavery was abolished in law as a consequence of a bloody war, but all that bloodshed did not
restore enough rights and respect to the negroes. "There is
no other way," he advised, "than the way of non-violence--
a way, however, not of the weak and ignorant but of the
strong and wise." This advice, eventually, was taken with
a measure of success by the late Martin Luther King, Jr.,
and by Corazon Aquino contending the presidency of the
Philippines in the controversial elections of February 1986,
as well as by their followers.

The very fundamental proposition of universal relevance
that Gandhi concerned himself with was amalgamating the
concepts of freedom with peace. Placed in an extended environment where the subject peoples were struggling to liberate themselves from debasing, dehumanizing foreign exploitation, he took an unequivocal stand that violence was no
path to emancipation, particularly because it could cause an
uncontrollable conflagration that would willy-nilly make for a
disturbed world moving precipitously to self-destruction. By
fusing together the urge for freedom with the obligation of
peace, Gandhi broke new ground both in political philosophy
and in the history of nations. He offered the technique of
satyagraha and an attainable objective, swaraj, to replace
violence among nations without overlooking or abandoning the
idea of conflict among humans. His actual experience in the
field--the many satyagrahas he led and inspired--not merely
sustained but steadily strengthened his conviction that violence locally, or war internationally, was no solution to the
dilemma of peace and freedom.

In sum, then, in the credo and practice of nonviolence
Gandhi saw a sure path to permanent world peace, and a
vindication and glorious victory for Indian nationalism. But

he also saw in it the method that national liberation move- ments the world over must unreservedly adopt for success, security, and strengthening of world peace. The method of India was "the exact equilibrium of war;"[84] its "non-violent revolution was not a programme of 'seizure of power', but it [was] a programme of transformation of relationships ending in transfer of power."[85]

This was the model to be emulated by subject peoples throughout the world not only for guaranteed success and subsequent stability, but also for contributing in a meaning- ful manner toward ensuring a durable peace. Freedom of the subject peoples was neither a myth nor a luxury; indeed, it was the very breath of life for them, as vitally necessary as freedom is for the individual. It must, however, be based on nonviolence, for then it could never be "a menace to the equal independence of other nations or individuals."[86]

Those struggling for freedom must, insisted Gandhi, maintain nonviolence as their goal and progress steadily to- ward it; the attainment of freedom must be in exact propor- tion to the attainment of nonviolence by that nation (or peo- ple such as in South Africa).[87] Indeed, deeply convinced as Gandhi was that real freedom could be won only through non- violence, he enjoined and encouraged the freedom fighters: "Keep the lamp of non-violence burning bright in the midst of the present impenetrable gloom. The truth of a few will count, the untruth of millions will vanish like chaff before a whiff of wind."[88]

Gandhi ardently believed freedom and peace to be inex- tricably linked. Not only were they equally important, but one could not last without the other. Peace, in his considered opinion, was both a prerequisite and a consequence of swaraj --rule of the self and self-government, particularly of India.

NOTES

1. The Collected Works of Mahatma Gandhi (New Delhi: Government of India, Publications Division), vol. 17, p. 439; vol. 19, p. 9; vol. 20, pp. 485-486; vol. 22, p. 104; vol. 35, pp. 456-457; vol. 47, p. 419.

2. Ibid., vol. 30, p. 462.

3. D.G. Tendulkar, Mahatma, vol. 6 (New Delhi: Govern- ment of India, Publications Division, 1960), vol. 6, p. 169; Collected Works, vol. 15, p. 310.

4. Tendulkar, Mahatma, vol. 7, p. 3.

5. Quotations until note 6 are in Collected Works, sequentially, vol. 22, p. 104; vol. 16, p. 259; vol. 25, p. 20; vol. 17, p. 439; vol. 16, pp. 306-307.

6. Harijan, vol. 6 (New York: Garland, 1973), p. 290.

7. Ibid., p. 282.

8. Pyarelal, Mahatma Gandhi, The Last Phase, vol. 1 (Ahmedabad: Navajivan, 1956), p. 117.

9. Collected Works, vol. 25, p. 424; vol. 17, pp. 439-441.

10. "I hold a new order to be impossible if the [Second World] War is fought to a finish or mutual exhaustion leads to a patched up peace," wrote Gandhi in July 1940. Harijan, vol. 8, p. 214.

11. Collected Works, vol. 15, p. 268.

12. Harijan, vol. 5, pp. 41-42; vol. 7, p. 332.

13. Tendulkar, Mahatma, vol. 7, p. 2.

14. Harijan vol. 11, pp. 85-86; Collected Works, vol. 27, p. 244.

15. Harijan, vol. 7, p. 89; vol. 6, p. 395; vol. 7, p. 272.

16. Collected Works, vol. 17, p. 435.

17. Ibid., vol. 35, p. 231; vol. 41, p. 154.

18. Harijan, vol. 6, p. 282.

19. Collected Works, vol. 20, p. 44.

20. Harijan, vol. 7, p. 288.

21. Collected Works, vol. 16, pp. 552-556.

22. Harijan, vol. 7, p. 89.

23. Collected Works, vol. 28, p. 441.

24. Harijan, Vol. 7, p. 89.

25. Collected Works, vol. 48, pp. 416, 441.

26. Ibid., vol. 15, pp. 104-105.

27. Ibid., vol. 33, p. 364.

28. Pyarelal, Last Phase, vol. 1, p. 118.

29. Tendulkar, Mahatma, vol. 7, p. 2.

30. Harijan, vol. 10, pp. 404, 412; vol. 11, p. 156.

31. Collected Works, vol. 48, pp. 416, 441.

32. Pyarelal, Last Phase, vol. 1, p. 120.

33. Harijan, vol. 7, pp. 272, 260.

34. Ibid., vol. 10, p. 91; also Delhi Diary (Ahmedabad: Navajivan), p. 160.

35. Harijan, vol. 11, p. 422.

36. Collected Works, vol. 48, p. 352.

37. Harijan, vol. 11, pp. 498, 509.

38. Vincent Sheean, Lead Kindly Light (London: Cassell, 1950), p. 201.

39. Collected Works, vol. 40, p. 365.

40. Ibid., vol. 31, p. 142.

41. Ibid., vol. 67, p. 76.

42. Ibid., vol. 31, p. 142.

43. Harijan, vol. 7, p. 235.

44. Collected Works, vol. 31, p. 143; vol. 29, footnotes on pp. 275-276; vol. 51, p. 178; vol. 31, p. 415; vol. 48, p. 393; vol. 35, p. 121; vol. 62, p. 285.

45. Pyarelal, Last Phase, vol. 2, p. 138.

46. Collected Works, vol. 61, p. 329.

47. Harijan, vol. 10, pp. 14, 52.

48. Ibid., vol. 11, p. 385.

49. Ibid., vol. 10, p. 134.

50. Ibid., p. 20.

51. Collected Works, vol. 53, p. 68; vol. 28, pp. 20-21.

52. Ibid., pp. 305-306; vol. 48, p. 369; vol. 35, p. 385.

53. Harijan, vol. 10, p. 389.

54. Substantiation of the argument, sequentially, in Collected Works, vol. 15, p. 103; vol. 40, pp. 365, 104; vol. 26, p. 279.

55. This is very ably discussed in Raghavan Iyer, The Moral and Political Thought of Mahatma Gandhi (Delhi: Oxford University Press, 1973), pp. 260ff.

56. Collected Works, vol. 28, p. 187.

57. Harijan, vol. 9, p. 156.

58. Ibid., vol. 8, p. 129; vol. 9, p. 156.

59. Collected Works, vol. 28, p. 190.

> One way to promote world peace is to help India to attian her own [freedom] through truth and non-violence (vol. 46, p. 216).
> Would that India . . . demonstrate to a world groaning under the curse of the sword that the spirit does triumph over the sword in national affairs as it has ever been shown to have triumphed in individual affairs (vol. 39, p. 460).
> [India's freedom through nonviolence] shall have made the largest contribution to the world peace (vol. 40, p. 167).
> There can be no real international peace so long as India remains fettered by a foreign power (vol 47, pp. 300, 390; vol. 40, p. 364).

60. Ibid., vol. 48, p. 401; vol. 32, p. 418; also Tendulkar, Mahatma, vol. 6, p. 11.

61. Collected Works, vol. 3, pp. 353-354; vol. 28, p. 191.

62. Pyarelal, Last Phase, vol. 1, p. 119.

63. Collected Works, vol. 23, p. 204.

64. Ibid., vol. 48, p. 401; vol. 28, p. 332.

65. Ibid., vol. 18, p. 134; vol. 48, p. 401.

66. Harijan, vol. 9, p. 302.

67. Pyarelal, Last Phase, vol. 1, pp. 118-119.

68. Harijan, vol. 8, p. 114; vol. 11, p. 177.

69. Ibid., vol. 10, pp. 14, 95.

70. Ibid., p. 116.

71. Ibid., vol. 11, p. 184, vol. 12, p. 513.

72. Collected Works, vol. 19, pp. 109, 12.

73. "If our present struggle remains non-violent to the very last, we shall win Swaraj without having incurred, during our whole struggle, the loss in men and money incurred in a single day of the European war [1914-18]" (ibid., vol. 43, pp. 381, 380).

74. Ibid., vol. 47, p. 390; vol. 41, pp. 153-154.

75. Ibid., vol. 47, p. 397.

76. Ibid., vol. 48, p. 182.

77. Ibid., vol. 28, p. 22; vol. 21, p. 516; vol. 47, p. 389; vol. 19, p. 12.

78. Harijan, vol. 9, pp. 197, 181.

79. Collected Works, vol. 46, pp. 143-144; vol. 35, p. 457; vol. 30, p. 462. "Freedom of India would demonstrate to all the exploited races of the earth that their freedom is very near" (Tendulkar, Mahatma, vol. 7, p. 2).

80. Ibid. Naturally, Gandhi considered foreign possessions like Goa on the Indian subcontinent to be absolutely integral parts of India and its people. For his concern and comment in this regard, see Harijan vol. 10, pp. 208, 235, 305; Collected Works, vol. 28, p. 364.

81. Collected Works, vol. 23, p. 201; vol. 30, p. 462; vol. 38, p. 256.

82. Ibid., vol. 17, p. 460; vol. 26, p. 140; vol. 48, p. 416.

83. Harijan, vol. 7, pp. 301-302; vol. 5, p. 41; vol. 9, p. 156.

84. Collected Works, vol. 48, p. 410.

85. Tendulkar, Mahatma, vol. 7, p. 29.

86. Collected Works, vol. 42, p. 437.

87. Harijan, vol. 10, pp. 134, 206-207.

88. Collected Works, vol. 3, p. 462; vol. 19, p. 390; vol. 27, p. 134.

7
A DURABLE PEACE

Humanity's urge for peace is innate and insatiable. Today, as never before, the urgency of securing it on a lasting basis is at once desperate and attainable. Gandhi's thought and action flowed from this deep rooted conviction. "Not to believe in the possibility of permanent peace," he held, was to "disbelieve in the Godliness of human nature. Methods hereto adopted have failed because rock bottom sincerity on the part of those who have striven has been lacking. Not that they have realised this lack."[1]

Sadly, the peoples of the world had never really or earnestly looked for the means conducive to the attainment of lasting peace. A dedicated band of war resisters emerged in Europe, of course, bending all their energies to combat the war fever, and it was fortuitous that their ranks were swelling and the futility of violence had widely come to be recognized there. However, that was not enough. And mere declarations, or signing the instruments, of peace by nations were of no use at all. The future, Gandhi was convinced, lay with those who believed in peace and peaceful methods, but, unfortunately, "we have not half explored these methods and their immense possibilities." The means adopted in this regard had all been wrong, as he indicates in the following statement.

> The attainment of real peace is impossible except for greater scientific precision, greater travail of the soul, greater patience and greater resources than required for the invention and consolidation of the means of mutual slaughter.
>
> It cannot be attained by a mere muster roll signed by millions of mankind desiring peace.

> But it can, if there is, be a few devoting
> themselves to the discovery of the means.
> Their effort being from within will not be
> showy but then it will not need a single
> farthing.[2]

As a beginning, all those who desire it should join together without reservations and the spell of specifics. But
most crucially, the call for peace will be a cry in the wilder-
ness so long as the spirit of nonviolence does not dominate
millions of men and women. It was nonviolence, and nonviolence alone, that in Gandhi's opinion constituted the sine qua
non of peace.

According to Gandhi, Europe thought in terms of violence,
which found due reflection in its programs and policies, because it believed in violence. Likewise, the whole structure of
the Soviet Union was based on force and violence. But democracy and violence, in Gandhi's belief, were just not compati-
ble. Apparently, the states that superficially or nominally
looked or pretended to be democratic must either become truly
and operationally democratic--possible only if they become
courageously nonviolent--or they must discard the hypocriti-
cal veneer and become frankly, honestly totalitarian.

Gandhi firmly believed that if nonviolence were adopted
as the design of life by an individual, or a nation, "it will
promote their happiness and peace and it would be their highest contribution to the attainment of the world peace." And
he considered it blasphemous to say that nonviolence could
be practiced only by individuals and not by nations; after all,
nations he pointed out are composed of individuals. All that
was required was reeducation, proper education, oriented
toward the goal of peace through nonviolence. This explains
his faith in UNESCO, which he considered to be the only
agency of the United Nations that could make a positive contribution toward lasting peace: "Real security and lasting
peace cannot be secured so long as extreme inequalities in
education and culture exist as they do among the nations of
the world. Light must be carried even to the remotest homes
in the less fortunate countries which are in comparative darkness. . . ."[3]

In unconditional acceptance of nonviolence, and its method, Gandhi saw the possibility of raising foundations pure
and strong for an enduring peace in the world. But how
could disputing nations resolve their differences? Gandhi's
firm answer was: through satyagraha. The technique of satyagraha was rooted in and permeated by truth. The differences
between nations arose over the perception of their interests.
The moral force of satyagraha would oblige a nation to ascertain its true interests, and in a spirit of cooperation rather
than selfishness or malice. Once that is done, interstate rela-

tions could come to be founded upon mutuality and accommo-
dation rather than competition, aggrandizement, and domination. After all, even today the nations do discover some
areas of cooperation and mutual benefit when they want to
and set seriously about doing so. One may cite the détente
of the 1960s and after as an example of what Gandhi may
have had in his mind. The choice of satyagraha would indeed
transform the very character of the state, and thus of interstate relations and the consequent state of the world. Non-
violence and satyagraha together will usher in a lasting peace.

Flowing from, and conditioned by, his philosophy of non-
violence, Gandhi's blueprint for a permanent peace included
earnest suggestions and strong support for the spirit of universal brotherhood, an attitude of internationalism, the promotion of world federation and world government, the absolute
necessity of total disarmament, and unilateral disarmament
even if others would not.

TRUE INTERNATIONALISM

Nations, according to Gandhi, must learn to live for each
other, and they are quite capable of doing so--even the defeated Germans: "all that is needed is a transmutation of their
marvellous energy for the promotion of the progress of the
world as a whole, rather than its application for their own
against that of the whole world."[4] He asserted that the great
powers assembled at San Francisco for the founding of the
United Nations ought to work for "parity among all nations--
the strongest and the weakest--the strong should be the
servants of the weak not their masters or exploiters."[5] Brotherhood of all peoples in the world, which is the ultimate goal
of humanity or at any rate should be, could only be based on
the plinth of their national freedoms, according to Gandhi:
"Hinduism insists on the brotherhood not only of all mankind
but of all that lives. . . . The moment we have restored real
living equality between man and man, we shall be able to establish equality between man and the whole creation. When
that day comes we shall have peace on earth. . . ."[6]

Evidently, then, Gandhi had no doubt whatsoever that
nationalism, instead of being narrow or exclusive, is a vital
prerequisite for internationalism. The passion with which he
made India's independence the linchpin of world peace exem-
plifies this belief. According to Gandhi, one could not be an
internationalist without being a nationalist. Internationalism
would come only if nationalism became a reality, that is, only
when peoples of different countries and cultures were organized in such a way that they were capable of acting as one
being.

Nationalism was not bad in itself, of course. It was the selfishness, the spirit of exploitation that inhered in those with power that, in Gandhi's view, impelled them to fight others and grow bigger at the latter's expense and ruin. He was emphatic that India's nationalism had broken new ground insofar as it was "seeking full self-expression for the benefit and service of humanity at large. I cannot possibly go wrong so long as I do not harm other nations in the act of serving my country."[7]

Even in the mid-1920s, Gandhi wanted "to think in terms of the whole world," for his patriotism include[d] the good of mankind in general."[8] He proclaimed, "My nationalism is intense internationalism." That is how, in his logic, hatred need not be an attribute of nationalism. Those who believed it was were to be pitied, for they were laboring under the grossest of delusions:

> So long as they retain that attitude the prog-
> ress of this country [India], the progress of
> the world is retarded. . . .
> The world is weary of it. We see the fatigue
> overcoming the Western nations. We see that
> this song of hate has not benefited humanity.
> Let it be the privilege of India to turn a new
> leaf and set a lesson to the world.

Gandhi's views of nationalism are expressed thus:

> Internationalism pre-supposes nationalism--not
> the narrow, selfish, greedy spirit that often
> passes under the name of nationalism that,
> whilst it insists upon its own freedom and
> growth, will disdain to attain them at the ex-
> pense of other nations.[9]

Apparently, in Gandhi's version of nationalism, there is a strong undercurrent of cooperation born of and necessitated by the understanding of one people by the others in a spirit of humility and sympathy rather than arrogance. He wants insolence and ignorance to be banished forever in the concerns and calculations of nations when it came to interacting with one another. Gandhi's nationalism was

> as broad as the universe. It includes in its
> sweep even the lower animals . . . all the na-
> tions of the earth. . . .
> If I possibly could convince the whole of
> India of the truth of the message, then India
> would be something to the whole world for which
> the world is longing.

> My nationalism includes the well being of the
> whole world. I do not want my India to rise on
> the ashes of other nations. I do not want India
> to exploit a single human being. I want India
> to become strong in order that she can infect
> the other nations also with her strength. Not so
> with other nations of the world, not so with a
> single nation in Europe today.[10]

The West, alas, was really "too materialistic, selfish and nar-
rowly nationalistic. What we want is an international mind,
embracing the welfare and spiritual advancement of all mankind."

Gandhi was certain that internationalism did not mean the
loss of national identity: "we do not want to follow the frog-
in-the-well policy, nor, in seeming to be international, lose
our roots. We cannot be international, if we lose our individ-
uality, i.e., nationality." Nations must promote nationalism,
of course, but such promotion should always be in the spirit
of universal brotherhood and should always "depend on the
rule [that the] stronger should help the poor."

Such a spirit, Gandhi had no doubt, could really and
quite easily be developed by means of positive nationalism:
unless people could serve their family and their village, they
could not serve the world. Nationalism had no malice, no ill-
will or contempt--it had only goodwill and peace in it. Unless
one learned to love one's neighbors, one could not possibly
cultivate the spirit of love for the rest of the world.

Gandhi described himself as a cosmopolitan. He could not
possibly be anything else, if he wanted his credo and gospel
to have universal relevance. And his mission was to unite the
world by "grouping unities":

> Unless I group unities I shall never be able to
> unite the whole world. Tolstoy once said that
> if we would but get off the backs of our neigh-
> bours the world would be quite all right without
> any further help from us.
>   And if we can only serve our immediate
> neighbour by ceasing to prey upon them, the
> circle of unities thus grouped in the right
> fashion will ever grow in circumference till at
> last it is co-terminus with that of the whole
> world. More than that it is not given to any
> man to try or achieve.[11]

This is what he meant by his axiom "as in the body so in the
universe." In this very spirit he readily agreed with Nehru
that

In these days of rapid intercommunication and
a growing consciousness of the oneness of all
mankind we must recognise that our nationalism
must be consistent with progressive internationalism. India cannot stand in isolation and unaf-
fected by what is going on in other parts of the
world. I can, therefore, go the whole length
with you and say that we should range ourselves with the progressive forces of the world.[12]

It is not surprising, therefore, to find Gandhi expressing
his revulsion against the idea of "Asia for Asians" as an anti-
European combine. Brimming with the vision of world unity,
he could not bear the thought of being a "frog in the well."
Takaoka, a member of the Japanese Parliament, sought a message from Gandhi at Sevagram for his party, which was just
founded with the motto "Asia for Asians." Gandhi's response
was blunt and firm:

I do not subscribe to the doctrine Asia for the
Asiatics, if it is meant as an anti-European
combination.
How can we have Asia for the Asiatics unless we are content to let Asia remain a frog
in the well? But Asia cannot afford to remain
a frog in the well.
I have a message for the whole world, if it
will only live up to it. There is the imprint of
Buddhistic influence on the whole of Asia, which
includes India, China, Japan, Burma, Ceylon
and the Malaya States. I said to the Burmese
and the Ceylonese that they were Buddhist in
name; India was Buddhist in reality. I would
say the same thing to China and Japan. But for
Asia to be not for Asia but for the whole world,
it has to learn the message of Buddha and deliver it to the world. Today it is being denied
everywhere.
Therefore . . . I have no message to give
you but this, that you must be true to your
ancient heritage. The message is 2,500 years
old, but it has not been truly lived. But what
are 2,500 years? They are but a speck in the
cycle of time. The flower of non-violence, which
seems to be withering away, has yet to come to
full bloom.[13]

Gandhi was unequivocal in his belief in world unity at the
Asian Relations Conference, held in Delhi on April 1, 1947:

Asian countries will strive to have only one
world. If you work with fixed determination,
there is no doubt that, in our generation, we
will certainly realise this dream. I will not like
to live in this world if it is not to be one. . . .
Uniting Asia to wage a war against Europe
or America is not India's mission. . . . I will
feel extremely sorry if India having won independence through essentially and predominantly
non-violent means was going to use their independence for the suppression of other parts of
the world.[14]

In Gandhi's view "for a non-violent person the whole world
is one family." If Gandhi had his way,

The post-war policy of the free National Government of India would be to promote commonwealth of all world states including, if possible
belligerent states also, so as to reduce to the
minimum the possibility of an armed conflict between different states.[15]

The League of Nations, he thought accordingly, could
not further the ideal of one world, for it was not universal.
A truly international League could exist only when all its
members, big or small, were independent, with the nature of
that independence corresponding to the extent of nonviolence
assimilated by the nations concerned. Gandhi is absolutely
clear about one thing: "In a society based on non-violence
the smallest nation will feel as tall as the tallest. The idea of
superiority and inferiority will be wholly obliterated."[16]
Just as Gandhi dismisses exclusivist, expansionist nationalism as undesirable and a cause of strife in the world, so he
would have little sympathy with supranationalism and organizations based on that principle. Although the idea of supranationalism facilitates relatively greater cooperation and possible understanding, it is not devoid of the proclivity to
dominate and exploit the rest.
In short, then, Gandhi pleads for liberal nationalism and
makes it the basis of his passionate internationalism, for
both, in his firm view, complement each other. And in his
concept of nationalism, swadeshi forms an integral, crucial
element.

THE IMPERATIVE OF SWADESHI

In a literal sense swadeshi means "home manufacture."
But the main thrust of Gandhi's advocacy and insistence of

swadeshi was on the need for self-reliance, which in turn
contributed to world peace. Seething poverty in India naturally moved Gandhi to conclude that independence did not
mean mere political emancipation from foreign rule. He believed that in order to be meaningful independence must have
a substantive economic content and criterion. Indeed, fiscal
autonomy and health were very vital for the life of a nation.
And that, he was confident, could be attained through swadeshi,[17] particularly in the case of India.

The history of nations, Gandhi said, bore testimony to
the fact that nations unable to preserve their economic independence for want of a policy of swadeshi had fallen, whereas those that could had always enjoyed swaraj. Citing the
example of the smaller European states of his time, he pointed
out that they survived as independent economic entities because they practiced swadeshi. Indeed, every nation that was
independent followed swadeshi in its own way. Switzerland
and Denmark, for instance, achieved this by maintaining
manufactures and trades that were suited to their own needs
and were free from outside interference. Swadeshi and swaraj,
in Gandhi's opinion, were thus inexhorably interdependent--
if not actually synonymous. The reign of foreign domination
would easily come to an end once the country ceased to be
dependent for its material requirements on others and became
instead capable of meeting its needs from its own resources.

> The sword will be sheathed as soon as Manches-
> ter calico ceases to be saleable in India. It is
> much more economical, expeditious and possible
> to give up the use of Manchester and, there-
> fore, foreign calico than to blunt the edge of
> Sir William's sword. The process will multiply
> the number of swords and, therefore, also miser-
> ies in the world.
>
> Like opium production, the world manufac-
> ture of swords needs to be restricted. The
> sword is probably responsible for more misery
> in the world than opium.
>
> Hence do I say that, if India takes to the
> spinning-wheel, she will contribute to the re-
> striction of armament and peace of the world as
> no other country and nothing else can.[18]

A fuller appreciation of Gandhi's position here calls for
some explanation. Clothing is one of our basic necessities.
Before the advent of colonialism in India, the economy of the
country was flourishing enough to meet the basic needs of
the people, leaving enough surpluses for export. That is
what had drawn the foreigner to India in the first place. The
village at that time was virtually a self-contained economic

unit that satisfied all the basic needs, and more, of its in-
habitants. Handicrafts and cottage industry flourished, and
agriculture supported a satisfying quality of life of the people in India.

Colonialism destroyed all that by policies coercively en-
forced to pursue its own interests. The compulsions and
requirements of industrialization at home obliged the colonial
rulers to supplant the prevailing economic system by a pattern that not only met their own needs but also aimed at
strengthening the shackles of the subject people by making
them dependent upon the metropolitan country, in this case,
England. As industrialization there grew, so did the dependence of India and its people on England.

The spinning jenny revolutionized weaving and cloth
making. Soon the textile mills of Manchester and Lancashire
were producing huge quantities of cloth that duly swamped
India, impoverishing in the process the craftworkers and
weavers in the villages along with the others. In reviving
and rejuvenating these sources, basic to the life and economy
of the village, Gandhi saw the only hope for the salvation of
the people of India. Hence his insistence on swadeshi, which
included not only cloth but all those items that could be--and
not long ago were--produced in the small, self-reliant Indian
village.

Gandhi thus argued that India's independence would be
hollow as long as it remained dependent for its basic needs
upon England. The industrial and corresponding technological
development of over a century in that country had increased
the gap tremendously with regard to India. Only an India relying upon its own resources and strength could possibly reduce or bridge that gap. Promoting and using swadeshi was
the obvious first step India must take for its independence
and giving meaning and substance to that independence.

This explains his call for stopping all imports from England so that India could regenerate its economy by its own
efforts. In his opinion, this would improve international relations, for India's relations then "will no longer be based on
consideration of self-interest but will be inspired by concern
for general welfare."

Gandhi's argument may not make much sense to many
modern economists, also because trade and commerce are major contributory factors to international interdependence and
seeming unification of the world. But due recognition of its
merit is reflected in the growing clamor for a new international economic order, as also in many an aspect of the North-South
dialogue, a discussion of which need not detain us here.

In any case, given the context of Gandhi's times, his
economic philosophy, and, most important, his priorities, his
argument for swadeshi and its integral component, self-reli-

ance, are perfectly consistent and very relevant. For instance, often he declared:

> Under my scheme . . . men incharge of machinery will think not of themselves or even of the nation to which they belong but of the whole human race. The Lancashire men will cease to use their machinery for exploiting India and other countries, but on the contrary they will devise means of enabling India to convert in her own villages her cotton into cloth. Nor will Americans, under my scheme, seek to enrich themselves by exploiting the other races of the earth through their inventive skill.[19]

The spinning wheel (charkha), which he held to be a "sign of peace," was crucial to Gandhi's concept of swadeshi, for it was through revival and resort to it that the rural economy in the country could be oriented toward self-reliance and finally to world peace. It was this belief that made him claim that the spinning-wheel was the most appropriate and effective answer to the atomic bomb. Andrew Freeman of the New York Post asked him if the spinning wheel had a cultural and therapeutic value for the malaise infecting the West, which had culminated in the atomic bomb, and if the wheel could serve as a counterweapon to the bomb. Gandhi responded that he did not have the "slightest doubt that the saving of India and of the world [lay] in the wheel."[20] He further argued that, in order to save the world from destruction, life had to be simplified; human dignity could be sustained only by serving everyone--even the last, the smallest, or seemingly most insignificant--on the earth: "We must do even unto this last [creature] as we would have the world do by us. All must have equal opportunity. Given the opportunity, every human being has the same possibility for spiritual growth. This is what the spinning-wheel symbolises."

Apparently, Gandhi's seminal ideas in this regard have much more than meets the eye, and it would not be productive to interpret his statements literally. The charkha for him has both literal and symbolic significance. Literally, it meets the immediate basic needs of the individual by producing the cloth through a fruitful use of one's time, which in any case is plentiful in the rural milieu; it precludes dependence upon the mercy or charity of others and enables one to meet a basic need in the dignity of one's own labor. On a symbolical level, the wheel signifies the spirit of self-reliance that sustains life. A complete discussion of his views on the value of the charkha, which he spelled out in his writings, is beyond the scope of this work, however.

The essential point to bear in mind here is that, by stressing the role of the charkha in the life of the individual, Gandhi is drawing attention to an honorable, dignified means of resisting exploitation and preventing alienation inherent in the entire system of industrial production. Industrialization in the world has occurred unevenly and at the cost of so much misery for the majority who feed the phenomenon by being merely suppliers of raw material and cheap or free labor. More often than not the industrial relations are marked by exploitation and consequent strife instead of equality and cooperation. Yet the charkha, Gandhi is absolutely certain, facilitates cooperation among millions in various ways. Through such cooperation "on the largest scale known to the world," he wants to teach "the uselessness, nay, the sinfulness of exploitation of those weaker than ourselves."[21]

Thus, he insists, a self-supporting, self-reliant India-- chiefly through swadeshi, of course--would be a proof against the temptation to exploitation. This, in turn, will make India least attractive to the greed or caprice of any power, Western or Eastern, and would automatically make India secure without its having to carry the burden of expensive, wasteful armaments. "Her internal economy [of swadeshi] will be India's strongest bulwark against aggression."[22] In this way, a sturdy contribution to a durable peace would naturally have been made, for self-support and self-reliance were the very essence of a peaceful world order.

THE COMPROMISE OF A WORLD
FEDERAL GOVERNMENT

Gandhi's belief in universal brotherhood, and his view of internationalism, quite logically led him to think in terms of and advocate a world government preceded, as a compromise, by a world federation based on nonviolence:

> Yes I claim to be a practical idealist. I believe in compromise so long as it does not involve the sacrifice of principles. I may not get a World Government that I want just now but if it is a government that would just touch my ideal I would accept it as a compromise.
>
> Therefore, although I am not enamoured of a world federation I shall be prepared to accept it if it is built on essentially non-violent basis.[23]

Nations must eventually unite. Gandhi abhored "isolated independence," least of all of India; rather, he espoused "voluntary interdependence."[24] Whereas isolated independence could easily menace the world, a federation of friendly interdependent states the world over--for which India was always ready --would unite the world in lasting peace. And if by India's efforts such a world federation of independent states comes into being, the hope of the kingdom of God, otherwise described as ramarajya, assured Gandhi, may legitimately be entertained.

Gandhi saw neither a contradiction between the freedom of India and a world federation brought about on a voluntary basis, nor a matter of choosing one to the rejection of the other. Shortly before his arrest in August 1942, Gandhi was asked, "Instead of striving for India's freedom why would you not strive for a far greater and nobler end--world federation? Surely that will automatically include India's freedom as the greater includes the less." Conceding that federation was "undoubtedly a greater and nobler end" than to be merely "self-centred, seeking only to preserve their own freedom," Gandhi pointed to the obvious fallacy implied in the question; namely, how could the subject nations be federated with the free as equals? So, clearly,

> The very first step to a world federation is to recognise the freedom of conquered and exploited nations. Thus, India and Africa have to be freed.
>
> The second step would be to announce to and assure the aggressor powers, in the present instance the Axis powers, that immediately [after] the war ends, they will be recognized as members of the world federation in the same sense as the Allies. This presupposes an agreement among the members of the world federation as to the irreducible fundamentals.[25]

The agreement on irreducible fundamentals was a necessity-- otherwise, the federation would fall to pieces under the slightest strain--and it must come voluntarily; nonviolence is the basis of voluntariness. Since, of all the nations of the world, India was the only one with the message of nonviolence, it must first have its immediate freedom to be able to play a constructive part in the creation of a world federation later on. According to Gandhi, India's freedom as a first step toward a world federation was necessary also because,

> If I can get freedom for India through nonviolent means, power of non-violence is firmly established, empire idea dissolves and the World State takes its place in which all the states of the world are free and equal, no state has its military.[26]

And, asserted Gandhi,

> A free democratic India will gladly associate
> herself with other free nations for mutual
> defence against aggression and for economic
> co-operation. She will work for the establish-
> ment of a real world order based on freedom
> and democracy, utilising the world's knowledge
> and resources for the progress and advance-
> ment of humanity.[27]

More than "a world federation of free and independent states," Gandhi considered it to be vitally necessary for these states to unite under one central governing body composed of their representatives. Indeed, "that is the only condition on which the world can live," as he told the correspondent of the United Press of America in June 1947. He viewed the world as an organic whole: "God has so ordered this world that no one can keep his goodness or badness exclusively to himself. The whole world is like the human body with its various members. Pain in one part must inevitably poison the whole system."[28] But this one world, as he told George Catlin, had to be founded upon respect for truth.[29] There was too much deception and hypocrisy about contemporary international relations. People declared themselves to be righteous, even in the face of utterly unjustified facts of murder and sudden death in the world; one must not compromise or come to terms with these ugly facts. The brightest prospect of world unity, Gandhi assured Catlin, rested upon respect for truth and ahimsa; nonviolence was the most appropriate and right course always, for nonviolence had never done any harm to anybody.

True to his faith of a lifetime, Gandhi, accordingly, pleaded for the adoption of nonviolence as a way of life by the nations of the world as "it will promote their happiness and peace," while, at the same time, being "their biggest contribution to the attainment of world peace after which we are all hankering."[30] As early as 1919, he had written in Young India,

> It is [the] law of love which, silently but
> surely, governs the family for the most part
> throughout the civilised world.
>   I feel that nations cannot be one in reality,
> nor can there be activities conducive to the
> common good of the whole humanity, unless
> there is this definition and acceptance of the
> law of the family in national and international
> affairs, in other words, on the political plat-

form. Nations can be called civilised only to
the extent they obey this law.

Thus a world state based on and permeated with nonviolence constitutes a prominent feature in Gandhi's blueprint
for permanent peace. But he was realistic enough not to anti-
cipate "a time in India or the world when all will be followers
of ahimsa," though he did "contemplate a time when in India
we shall rely less on brute force and more on soul force."[31]
A state without police was very much in his scheme of
things, of course, and its attainment quite feasible too, but
Gandhi also conceded that it requires a higher degree of
courage and unity than is ordinarily available. Therefore,
"whilst I can invite all states to do without police or army,"
he candidly admitted, "I have not yet been able to bring myself to believe that you can preserve a society without police.
. . . You can thus say that my toleration of police is a limi-
tation of non-violence." And "police there will be even in
satyayuga [the age of truth and perfection]," he felt.
Accordingly, Gandhi envisaged that as a transitory measure there "may be a world police to keep order in the absence of universal belief in non-violence."[32] He would allow
"an armed police to enforce the lightest terms of peace. Even
this retention of an international police," said Gandhi rue-
fully, "will be a concession of human weakness not by any
means an emblem of peace."[33]
But never for a moment did Gandhi see any need for an
army for, in his opinion, an army was "opposed to non-violence." Whatever tolerance or sympathy he might have had
for it in his early days in South Africa for its virtues of discipline and devotion to duty had dwindled drastically, if not
completely evaporated. He is frank, however, in acknowledging,

> It is my inability to convince my people to do
> without army. I have not mustered sufficient
> strength to pit non-violence against thieves
> and scoundrels and cut-throats but I can ask
> people to pit non-violence against hordes of
> the army.
> If perchance India wins her deliverance
> through non-violence we may perhaps show to
> the world that it is not necessary to have an
> army state.
> I do not regard it utopian to think of a
> state without army, but it requires a higher
> degree of courage and purity.[34]

He might well have added that, until humanity regenerates the high moral fiber appropriate to such a transformation, an army would be necessary.

The nearest Gandhi gets to the idea of an army is that of a nonviolent "army," which differed fundamentally from the usual violence-oriented organization. The difference lay in the nature of the discipline his nonviolent army would observe and emulate: the discipline would come from within the "soldier" rather than from without. The soldier of the usual army "yields obedience whilst at war, but will yield to wild licence when free from it. But a non-violent soldier carries discipline in his heart and will carry an atmosphere of restraint in every walk of life."

It appears that for establishing peace Gandhi had far greater hope in the British Commonwealth of Nations than in the United Nations. But the Commonwealth must first be transformed into "a fellowship of free nations joined together by the 'silver cords of love'. . . . I would have India enter freely into such a fellowship and with the same rights of equality for Indians as for other members of the Commonwealth." He explained that

> my ambition is much higher than independence. Through the deliverance of India I seek to deliver the so-called [weak and backward] races of the earth from the crushing heels of western exploitation in which England can have the privilege of becoming a partner if she chooses.[35]

He elaborated the thought by reiterating what he had always maintained:

> Purna Swaraj [complete independence] does not exclude association with any nation--much less with England. But it can only mean association for mutual benefit and at will. Thus, there are countries which are said to be independent but which have no Purna Swaraj, for example Nepal.[36]

To that example he could have added the countries of central and southern America, where the type of independence Gandhi visualized was far from being a reality or even an aspiration. Anyway, when eventually India became a full-fledged and equal member of the transformed Commonwealth of its own volition, Gandhi considered it to be an event of great importance, for the members had willingly launched a multiracial Commonwealth. This, in his opinion, augured well for the forces of peace in the world, for justice had at last been done to a people who had long clamored and struggled for it.

THE DECEPTION OF DISARMAMENT

For years, Gandhi campaigned for the view that peace was a matter of justice rather than of strength: "all the world over a true peace depends not upon gun powder but upon pure justice." Unless nations learned to be just by renouncing exploitation and their total reliance on force, there could never be peace on the earth. Armaments were both a symptom of force and a temptation to use force. Thus, instead of talking of a mere restriction of armaments, their quantity and quality, Gandhi pleaded passionately for complete, universal disarmament.

Apparently in thorough approval, he published in the columns of Young India, which he edited, an excerpt from a monthly called International Sunbeam:

> Total world disarmament, the only material safeguard for peace, should be the outward and visible sign of that inward mental disarmament on which alone outward peace can secure. So long, however, as one people is actually subjecting another to itself by superior military might even the very first step toward this inward mental disarmament has not been taken.[37]

Europe should take this first step, held Gandhi, "unless Europe is to commit suicide." The tendency to outdo one another in building up armaments would subside once that step were taken.

By the same token, it would be equally disastrous if India, after attaining its freedom, were to join the arms race: "For India to enter into the arms race is to court suicide. With the loss of India to non-violence the last hope of the world will be gone. . . ." Gandhi hoped "that India will make non-violence her creed, preserve man's dignity, and prevent him from reverting to the type from which he is supposed to have raised himself."[38]

The bloodshed of the Hindus and Muslims in the wake of partition and independence of India greatly distressed Gandhi, of course, but what exasperated him even more was the country

> swearing by the military and all that naked physical force implies. Our statesmen have for over two generations declaimed against the

> heavy expenditure on armaments under the
> British regime, but now that freedom from po-
> litical serfdom has come, our military expendi-
> ture has increased and still threatens to in-
> crease and of this we are proud!
>
> However . . . the hope lingers in me and
> many others that India shall survive this
> death dance and occupy the moral height that
> should belong to her after the training, how-
> ever imperfect, in non-violence for an un-
> broken period of thirtytwo years since 1915
> [the year Congress accepted the method of
> nonviolent struggle for the country's free-
> dom].[39]

Gandhi never reconciled himself to the partition of India. Its
aftermath of communal massacres in the Panjab and Bengal
saddened him further. Talk of strengthening and relying upon
the military as a means of meeting the situation, which was
surcharged with suspicion, anger, and frustration, grieved
him greatly.

He saw little sense in India after attaining its freedom
incurring a heavy defense budget and felt that the country
was groaning under this unnecessary and unsupportable burden: "We are convinced that we do not need the arms that
India is carrying."[40] He stuck to his oft-repeated position
that nothing could be more disastrous than India trying to
imitate or rival the Western powers, including the United
States, in their defense strategy based on military capability.

Gandhi unequivocally advocated that India--even Asia--
take a lead with regard to disarmament. At the concluding
session of the Asian Relations Conference at Delhi on April 2,
1947, he declared that the West

> is despairing of a multiplication of the atom
> bomb, because atom bombs mean utter destruc-
> tion not merely of the West but of the whole
> world, as if the prophesy of the bible is going
> to be fulfilled and there is to be a perfect de-
> luge.
>
> It is up to you to tell the world of its
> wickedness and sin that is the heritage your
> teachers and my teachers have taught Asia.[41]

His anguish at the talk in India of continuing with the military,
in fact strengthening it, is therefore in character.

Gandhi was certainly opposed, however, to the forced
disarmament of a nation. For instance, referring to the disarming of India during World War I, Gandhi declared with vehemence:

> Much as I abhor the possession or the use of
> arms, I cannot reconcile myself to forcible pro-
> hibition.
>
> As I said three years ago, this forcible dis-
> armament of a people will be regarded by his-
> tory as one of the blackest sins committed by
> the British Government against India. If people
> want to possess arms they ought to have them
> without ado. . . . We cannot learn discipline
> by compulsion. We must learn not to use arms
> or to use them with responsibility and self-re-
> straint, notwithstanding the right to possess
> them.[42]

In the same spirit, he condemned the disarming of the vanquished as a punitive measure. This was not merely due to
compassion; Gandhi believed that guilt of war was so pervasive that it was virtually impossible, and wrong, to apportion
all the blame for it to just one of the parties at the expense
of the other.[43]

If the great powers disarmed themselves, felt Gandhi,
they would not only escape the ravages of war but also cover
themselves in glory, restore sanity to the world, and earn
the eternal gratitude of posterity. These powers would, at
the same time, have to give up their imperialistic designs and
exploitation of the weak and hapless, while revising their own
way of life. All this, evidently, amounts virtually to a total
revolution, the only alternative to which, in Gandhi's opinion,
was absolute disaster. The great powers could not be expected to move in the direction of total revolution--contrary
to what lifestyle they had thus far been used to--spontane-
ously, of course.[44]

Nor could disarmament be worked out overnight; it would
take time. The one thing Gandhi was absolutely certain of
was that it would come only through the adoption of nonviolence by a nation--indeed, by all the nations of the world.
The evils of armaments can be cured, held Gandhi,

> by non-violence, which will eventually be the
> weapon of all nations. I say 'eventually' de-
> liberately, because we shall have war and arma-
> ments for a very long time. It is two thousand
> years since Christ preached his "Sermon on the
> Mount" and the world has adopted only a frag-
> ment of the imperishable lofty precepts therein
> enunciated for the conduct of man toward man.
> Until we take all Christ's principles to our
> hearts, war, hatred and violence will continue.[45]

It was wrong to think that armaments implied or imparted strength, for real strength, according to Gandhi, came by self-sacrifice, from within and not through physical force. He pointed out that the rishis (sages) of yore, themselves great warriors, realized the utter uselessness of force and "taught a weary world that its salvation lay not through violence but through non-violence."

Gandhi was emphatic that nations should strengthen themselves spiritually, for "Internal spiritual forces are stronger and induce a more certain and lasting life. It is not by arming yourself that you will guarantee peace to the world. External arms, guns, cannons and gas have only evil and passing results."[46] To say that one is arming oneself for self-defense was, he said, "a wretched plea . . . a bad thing," for, in effect, the result invariably was that you organized yourself to "prey upon ill organized communities and nations." The crux of the matter was that real disarmament could come only when "the nations of world cease to exploit one another."

Gandhi would not accept the proposition that, since disarmament chiefly depended upon the great powers, small, neutral and nonaggressive countries like Switzerland, for instance, should be forced to disarm. Gandhi asserted that the very fact of neutrality and nonaggressiveness of Switzerland rendered the army there completely redundant. Indeed, it would be a far superior thing for Switzerland to give the world a lesson in disarmament and establish that the Swiss are brave enough to live without an army. Nor was he willing to accept the proposition that the mere presence of the Swiss army had saved that country from being overrun by foreign armies. Rejecting both the deterrent role of the Swiss military forces as well as the value of military development, Gandhi astutely pointed out the fallacies involved:

> Will the questioner forgive me if I say that a double ignorance underlies this question?
> He deplores the fact that, if you give up the profession of soldiering, you will miss the education you receive in service and sacrifice.
> None need run away with the idea that because you avoid military conscription you are not in for a conscription of a severer and nobler type. When I spoke to you about labour, I told you that labour ought to assimilate all the noble qualities of soldiering: endurance and defiance of death and sacrifice. When you disarm yourself, it does not mean that you will have a merry time. It is not that you are absolved from the duty of serving your homes when you give up soldiering; on the contrary,

your women and children would be taking
part in defending your homes.

Again I am not talking to you without experience. In the little institution [sic] that we
are conducting, we are teaching our women
and children also how to save that institution
--as we are living among thieves and robbers.

Everything becomes simple and easy the
moment you learn to give up your own life in
order to save the life of others.

And lastly, it is really forgotten that safety
which an individual derives from innocence is
safety which no amount of arms will give you.

The second part of the ignorance lies in
the second part of the question.

I must respectfully deny the truth of the
statement that the presence of the Swiss army
prevented the War from affecting Switzerland.
Although Belgium had its own army, it was not
saved and if the rival armies had wanted a
passage through Switzerland, believe me, they
would have fought you also. You might have
fought in turn, but you would have fought
much better non-violently.[47]

Gandhi was also not deluded by the description of an
arms race among the powers of Europe, and elsewhere, as a
manifestation of their desire to eliminate war.[48] This was a
misconception at best and plain hypocrisy at worst for the
noble ideal of peace could not possibly be achieved by the
disastrous means of war. And where was even the slightest
indication that in acquiring more and more weapons the concerned nation had abjured the motive of being dominant and
jockeying for the position of strength? In his message to the
editor of a U.S. magazine, The Cosmopolitan, Gandhi wrote:

If recognised leaders of mankind who have
control over engines of destruction were
wholly to renounce their use with full knowledge of implications, permanent peace can be
obtained.

This is clearly impossible without the great
powers of the earth renouncing their imperialistic designs. This again seems impossible
without these great nations ceasing to believe
in soul destroying competition and the desire
to multiply wants and therefore increase their
material possessions.[49]

With regard to nuclear weapons, Gandhi thought it quite futile merely to restrict them; he seriously doubted that the nations possessing them would remain content merely with their possession. Dismissing the efficacy or permanence of "balance of terror" or "balance of nonuse" and the supposition that their "overkill" capacity by itself would induce and enforce nonviolence in the world, he emphasized that so long as the thinking in these quarters remained rooted in violence we will continue heading toward disaster and self-destruction. Indeed, "the violent man's eyes would [be] lit up with the prospect of much greater amount of destruction and death which he could now wreak," Gandhi said. At best the destructive capability of these weapons may temporarily postpone the outbreak of war. "Like a man gutting himself with dainties to the point of nausea and turning away from them only to return with redoubled zeal after the effect of nausea is well over. Precisely in the same manner will the world return to violence with renewed zeal after the effect of disgust is worn out."[50] Gandhi quite frankly considered the "balance of terror" to be nothing more than the "preparations for a third World War." He was willing to concede that atomic energy "may be utilised . . . for humanitarian purposes" but was sure that it would not be confined to that alone.

According to Gandhi, nations armed themselves out of fear of each other and to guard their imperialist possessions. Quite early in his life, echoing the philosophy of the Gita, Gandhi in his Hind Swaraj asserted that nations armed themselves because they were filled with the fear that others may take away their possessions; force was used when people were under the spell of fear, and "what is gained through fear is retained only for as long as fear is present."[51] At the same time, fear breeds hatred. Apparently then, the very first step that nations must take toward ensuring peace was to dispel fear and, along with it, mistrust. Gandhi said his instincts told him that if you do not intend violence to anyone, no one will use it against you either: "It is only when we are afraid of our opponent, we employ unclean strength like his that we learn unclean ways and so become weak. . . . If we meet uncleanliness with cleanliness, the total result would be less of uncleanliness and the people, the world, would be happier for this."[52]

In sum, then, Gandhi put it succinctly and without any equivocation when he declared:

> Peace will never come until the Great Powers courageously decide to disarm themselves. . . .
> I have an implicit faith--a faith that today burns brighter than ever, after half a century's experience of unbroken practice of non-violence

--that mankind can only be saved through non-violence.[53]

PERSUADING NATIONS TO UNILATERAL DISARMAMENT

Speaking about Britain, before World War II broke out, Gandhi stated:

> Someone has to rise in England, with the living faith to say that England, whatever happens, shall not use arms. They are a nation fully armed, and if they having the power fully deliberately refuse to use arms, theirs will be the first example of christianity in active practice on a mass scale. That will be a real miracle.[54]

He thus called upon the greatest nation of its times to make the noble beginning.

A realist and ever a man of action, Gandhi fully realized that, before general disarmament prevailed, a beginning had to be made somewhere.

> Some nation will have to dare disarm herself and take large risks. The level of non-violence in that nation, if that event happily comes to pass, will naturally have risen so high as to command universal respect. Her judgment will be unerring, her decisions will be firm, her capacity for heroic self-sacrifice will be great, and she will want to live as much for other nations as for herself.[55]

In calling upon Britain to take the initiative in unilaterally disarming itself, Gandhi seemed to be saying that a voluntary renunciation of its formidable military capability would make a tremendous and multifaceted impact on the nations of the world. His invitation also implied in a way the great faith he had in the people of that country--once they decided to do something, they could and did it well. And would it not be some atonement for their past sins?

All good things, Gandhi believed, began always with one single person. Thus, he urged that an initiative toward disarmament be taken soon by some country. To one of his biographers, Frederick B. Fisher (author of The Strange Little Brown Man Gandhi), Gandhi sent a message on September 31, 1931, for American Christians on World Peace and Disarmament in which he unambiguously stressed that

> Peace and disarmament are not a matter of re-
> ciprocity.
>   When real Peace and Disarmament come, they
> will be initiated by a strong nation like America
> irrespective of the consent and co-operation of
> other nations. . . .
>   As Thoreau has said so well, "all reform all
> the world over always began with one person
> taking it up". . . .[56] [Emphasis added]

Here is the very kernel of Gandhi's entire thinking and
campaign with regard to disarmament. Having concluded that
disarmament was moral, eminently desirable, and urgent in a
world quickly moving toward self-annihilation, the karmayogi
in Gandhi was moved to seek courses of action. In unilateral
disarmament he saw a bold but perfectly feasible course of
action in concrete terms. And he pressed for it whenever he
could relax the focus of his campaign for nonviolence and ramp-
ant international exploitation, particularly in the colonial world.

For much of his later life, Gandhi was preoccupied with
the particular issues raised by India's freedom struggle.
Since he had been thoroughly disillusioned about the intentions of the British Empire, he could now more easily explore
the implications of a total rejection of war. World War II re-
inforced this position both because of the urgent need to
consider what action to take if the Japanese invaded and
because of the awareness that war was now something that
involved a new level of destructiveness. As a result, what
emerges is a crystallized conviction of war as an evil that
must be prevented by eliminating its root causes and resorting to moral means in preference to arms or their regulated
use.

Asserting that "physical possession of arms is the least
necessity of the brave," Gandhi stated that nonviolence requires the strength and courage to suffer without retaliation,
to receive blows without returning any. Nonviolence, the
surest means of ushering in permanent peace, becomes futile
"unless the root cause is dealt with, and the root cause is
the greed of nations. If there were no greed, there would be
no occasion for armaments: the principle of non-violence
necessitates complete abstention from exploitation in any form.
Immediately the spirit of exploitation is gone, armaments will
be felt as an unbearable burden." He declared, "I have no
doubt that unless big nations shed their desire of exploitation and the spirit of which war is the natural expression
and atom bomb the inevitable consequence, there is no hope
for peace in the world."[57] In sum, an eternally peaceful
world federation could be raised only on the sturdy and deep
foundations of nonviolence; violence will have to be given up
in all shape and form in world affairs; and disarmament--a

contributory condition of such peace--will come only when nations shed their exploitation and the fetish for arms.

Certain that maneuvers like those of President Wilson
would not bear the fruit of peace expected of them since they
were rooted in deep hypocrisy, Gandhi ceaselessly endeavored to break a new path. Wilson had remarked, "After all,
if this endeavour of ours to arrive at peace fails, we have
got our armaments to fall back upon." Gandhi's rejoinder to
such an attitude was: "I want to reverse that position and
say 'our armaments have failed already. Let us now be in
search of something new, and let us try the force of love
and God which is Truth.'"[58] He wanted to show that physical
force "is nothing compared to the moral force and that moral
force never fails."[59] Moreover, not for a moment did he think
that peace could ever be attained in a piecemeal manner:
"Peace is unattainable by part performances of conditions
even as chemical combination is impossible without complete
fulfillment of conditions of attainment thereof."[60]

Real, effective, lasting peace, Gandhi reiterated ad in-
finitum, could only be attained through the rule of the "law
of love" saturating the life and style of nations the world
over. Love for one's own nation is not enough, for "such a
love is an armed peace." War will only be eliminated when the
conscience of humanity becomes elevated enough to recognize
the undisputed supremacy of the law of love in all walks of
life.[61] This was not all that utopian, or beyond the reach of
humanity, as might be imagined; Gandhi was absolutely confident that all this will come to pass here and perhaps in not
too distant a future. So the columns of Young India of June
23, 1919, articulated:

> Till a new energy is harnessed and put on
> wheels, the captains of older energies will
> treat the innovation as theoretical, impractical,
> idealistic and so on.
>   It may take long to lay the wires of inter-
> national love, but the sanction of interna-
> tional non-co-operation in preference to con-
> tinued physical compulsion . . . is a distinct
> progress towards the ultimate and real solution
> [p. 51].

And here, Gandhi saw an independent India in the role
of a catalyst. In an interview for the News Chronicle, soon
after his release from prison in 1944, he observed:

> I am a lover of peace through and through.
> After independence was assured, I would
> probably cease to function as adviser to the
> Congress and as an all war resister I would

have to stand aside, but I shall not offer
any resistance against the national Government or the Congress [if they decided to
support the war effort]. My co-operation
[to the war effort] will be abstention from
interfering with even tenor of life in India.
I shall work with the hope that my influence will be felt to keep India peace
minded and so affect the world policy towards real peace and brotherhood among
all without the distinction of race and
colour.[62]

Gandhi maintains, "perfect peace comes when mind and
heart are pure."[63] All his life, he bent the entire energies
of his being in purifying the heart and elevating the mind
of his countrymen with the contagion of his tireless pursuit
of truth--concretized in nonviolence, fearlessness, sense of
responsibility and duty, and a loving concern for his fellow
beings. Congenital optimist perhaps, but certainly one who
had implicit, unbounded, and a proud faith in his country
and countrymen, Gandhi seemed to expect unilateral disarmament in India. Perception by a newly independent India may
be blunted for a time with the euphoria of emancipation. But
Gandhi was confident that "even in my absence my influence
for peace will last; though I may be far away, my spirit will
remain behind."[64]

NOTES

1. The Collected Works of Mahatma Gandhi (New Delhi:
Government of India, Publications Division), vol. 62, p. 175;
vol. 28, p. 335; vol. 67, pp. 123, 420.

2. Ibid., vol. 66, pp. 73, 266; vol. 47, p. 390; vol. 31,
p. 141; Harijan (New York: Garland, 1973), vol. 6, p. 328;
vol. 11, pp. 412-413.

3. Collected Works, vol. 62, p. 174.

4. Ibid., vol. 30, p. 50.

5. Pyarelal, Mahatma Gandhi, The Last Phase, vol. 1
(Ahmedabad: Navajivan, 1956), p. 119.

6. Collected Works, vol. 62, p. 285.

7. Ibid., vol. 27, pp. 255-256; vol. 28, p. 186; vol. 31,
p. 45.

8. Ibid., vol. 28, pp. 126-127.

9. Ibid., vol. 31, p. 181.

10. Ibid., vol. 28, pp. 22-23, 134; vol. 45, p. 319; vol. 47, p. 323; vol. 35, p. 364.

11. Ibid., vol. 17, pp. 407-408.

12. Ibid., vol. 55, pp. 426-430.

13. Harijan, vol. 6, p. 404.

14. Ibid., vol. 11, pp. 113-114; vol. 8, p. 214.

15. Pyarelal, Last Phase, vol. 1, p. 30.

16. Harijan, vol. 7, p. 8.

17. Collected Works, vol. 15, pp. 486-490.

18. Ibid., vol. 28, p. 454; vol. 15, p. 376.

19. Ibid., vol. 28, p. 189; vol. 21, p. 161.

20. Harijan, vol. 10, p. 404.

21. Collected Works, vol. 28, p. 441.

22. Ibid., vol. 47, p. 90. Gandhi states, "Free India can have no enemy. And if her people have learnt the art of saying resolutely 'no' and acting up to it, I dare say, no one would want to invade her. Our economy would be so modelled as to prove no temptation for the exploiter." Harijan, vol. 7, p. 304.

23. Pyarelal, Last Phase, vol. 1, p. 120.

24. Collected Works, vol. 24, p. 395; vol. 25, pp. 481-482; vol. 47, p. 389; Harijan, vol. 10, p. 235.

25. Harijan, vol. 9, p. 265.

26. Pyarelal, Last Phase, vol. 1, p. 120.

27. Harijan, vol. 7, p. 278; vol. 9, p. 235.

28. Ibid., vol. 11, pp. 184, 154.

29. George Catlin, In the Path of the Mahatma (London: MacDonald, 1948), p. 281.

30. Collected Works, vol. 62, p. 174.

31. Ibid., vol. 2, p. 130; vol. 48, p. 158.

32. Pyarelal, Last Phase, vol. 1, p. 120.

33. D.G. Tendulkar, Mahatma, vol. 7 (New Delhi: Government of India, Publications Division, 1960), p. 3.

34. Collected Works, vol. 48, p. 158; vol. 47, pp. 277, 419; vol. 23, p. 106; vol. 19, p. 444.

35. Ibid., vol. 35, p. 457.

36. Ibid., vol. 45, p. 264.

37. Ibid., vol. 38, pp. 160-161; vol. 15, p. 285; vol. 28, p. 305.

38. Harijan, vol. 7, p. 305.

39. Ibid., vol. 11, p. 453.

40. Collected Works, vol. 48, pp. 200-201; vol. 41, p. 310.

41. Harijan, vol. 11, p. 117.

42. Collected Works, vol. 20, p. 406; vol. 27, p. 51.

43. Commenting upon the treatment meted out to the "war criminals," Gandhi said: "What is a war criminal? Was not war itself a crime against God and humanity and, therefore, were not all those who sanctioned, engineered, and conducted wars, war criminals? War criminals are not confined to the Axis Powers alone. Roosevelt and Churchill are no less war criminals than Hitler and Mussolini. . . . England, America and Russia have all of them, got their hands dyed more or less red--not merely Germany and Japan." Pyarelal, Last Phase, vol. 1, p. 114.

44. Harijan, vol. 6, p. 328.

45. Collected Works, vol. 45, p. 319; vol. 18, p. 133.

46. Ibid., vol. 48, pp. 393, 402-403, 441, 86.

47. Ibid., pp. 419-420.

48. Ibid., vol. 34, p. 318.

49. Harijan, vol. 4, p. 109.

50. Ibid., vol. 10, pp. 212, 197, 140, 217.

51. Mahatma Gandhi, Hind Swaraj (Ahmedabad: Navajivan, 1962), chap. 16.

52. Collected Works, vol. 19, p. 10.

53. Harijan, vol. 6, p. 395.

54. Collected Works, vol. 67, p. 76.

55. Ibid., vol. 48, p. 85.

56. Ibid., vol. 54, p. 335.

57. Harijan, vol. 3, p. 276; vol. 10, p. 389; Collected Works, vol. 49, p. 441; Gandhi's Correspondence with the Government 1942-44 (Ahmedabad: Navajivan, 1959), p. 143.

58. Collected Works, vol. 28, p. 23.

59. Ibid., vol. 15, p. 142.

60. Ibid., vol. 62, p. 175.

61. Ibid., vol. 31, p. 143.

62. Tendulkar, Mahatma, vol. 6, p. 256.

63. Collected Works, vol. 19, p. 10.

64. Ibid., vol. 48, p. 353.

8
THE MAHATMA'S MESSAGE OF PEACE

Essentially a man of his milieu and times who looked far beyond, Mohandas Karamchand Gandhi was not shaped by formal education or the influence of the famous and celebrated as much as is often claimed. The basic traits and aspects of his personality, thinking, and style of action had been defined by sources exclusively traditional, Indian--more Hindu than Jain or any other. On to these his genius grafted essentials from Christianity and other religions. In his endeavors at synthesis, he found encouraging corroboration, support in certain persons of eminence and their writings.

On war and peace, without a shadow of doubt, Gandhi's thinking and attitude were rooted in the Gita into which he read simultaneously the message of discipline and duty along with that of fearlessness, courage, and nonviolence. By exhorting ceaseless, selfless action in thought and deed, the Gita led Gandhi to abhor passivity in any shape or form and made him a karmayogi. These values and virtues combined seemed to constitute the path to truth, the sole concern of his life. Dedication to duty accounts for Gandhi's active association with certain violent conflicts including World War I, as indeed, it explains his decision to struggle for the emancipation of his countrymen. Admiration for the values of fearlessness, courage, discipline, and duty made him respect the soldier in the Boer War and inspired him to instill these same qualities in his countrymen. Gandhi's whole life thus became a saga of tirelessly striving to inculcate these values in his people by scrupulously practicing them himself.

The seeming ambiguity between his overarching, all-pervasive nonviolence on the one hand and his association with certain wars on the other is no contradiction at all. There is continuity and basic coherence in his approach; his early and later, mature positions are transparently consecutive.

Gandhi himself adhered to nonviolence with rare passion and sincerity, and his preference instinctively was for non-violence. But he was realist enough to realize human limitations: those not trained in nonviolence or who had not yet cultivated or perfected its technique may use violence in self-defense--for example, the reaction of the Jews and Poles in World War II or the question of Kashmir--rather than submit meekly. He also understood violence in certain other conditions; moreover, he took due cognizance of the fact that violence in life was unavoidable. He would be among the first to realize that theory and practice hardly ever coincide even as Euclid's line in practice never exactly coincides with his theoretical definition.

No greater injustice or injury can be done to Gandhi than to impose upon him the interpretation that he talked to and preached nonviolence in absolute terms or in perfect conditions when life itself is qualified in several ways at every step. Truth and realism lie in recognizing the fact of relativism rather than pushing life into the straight jacket of absolutism. As a perceptive scholar of Gandhi notes:

> It has more or less become a dogma, among students of Gandhian thought, to assume that violence and nonviolence are opposites which cancel each other out. They have tended to look upon the world of reality in terms of black and white, forgetful of the infinite nuances of grey that confront us at every turn. Life is not . . . a clear-cut progression from the infra-red of brute violence to the ultra-violet of ethereal nonviolence.
>
> Life is rather like a close-knit web, in which violence and nonviolence interpenetrate one another, not as contradictory elements but as complementaries. This interpenetration of seeming opposites--good and evil, light and darkness, life and death, violence and nonviolence--is the essence of reality. . . .[1]

Gandhi's nonviolence did not mean simply noninjury or nonkilling, not passive bodily action, but an inward disposition manifesting itself in positive activity. It has also to be remembered that his experiments led Gandhi from truth to truth. The seeming inconsistency, under the circumstances, becomes merely a reflection of the growth of his ideas. So he maintained that his later statements and position on any issue be taken as final. Gandhi was willing to compromise on what he considered to be inessentials and even conceded inconsistencies, but this did not invalidate his thorough condemnation of war and earnestness for peace in the world.

Additionally, his debate with the Western pacifists and war resisters illumined the fact that Gandhi rejected the Christian concept of a "just war," which condoned the violence of war in certain circumstances according to St. Augustine and Thomas Acquinas. His position was closer to that of Immanuel Kant, who held that there is no such thing as a just war. And yet, Gandhi did believe that of the two parties locked in violent conflict one of them had justice on its side. He wished that these pacifists did not shirk the dilemma of having to choose which of the two had justice on its side. Having decided conscientiously and objectively, they should lend their wholehearted nonviolent support to the side that was morally in the right. At the same time, they should persuade this side to offer nonviolent resistance in pursuit of its objectives rather than the violence of war.

By the time of World War II, Gandhi's ideas on war and peace had taken a finality lacking earlier. Nothing would now induce him to assist or associate in war, "an unmitigated evil." Accordingly, he tried to induce Congress to boycott the war totally, not to offer any help--conditional or unconditional--in the war effort. Insofar as he was satisfied that the Allied cause was "just," Gandhi was inclined to let the Congress offer them its moral support, at best. Or a more cynical view could be that he placed more hope on the Allies for India's independence, as evidenced by his reluctant tolerance of U.S. troops on the Indian soil. But this latter view does not seem to be tenable in the face of Gandhi's unswerving faith and conviction that India's emancipation would come entirely from its own strength and efforts, nonviolently.

In the event of foreign aggression on India, Gandhi's style of meeting it would be to mobilize the country's masses so completely that in the face of their flawless nonviolent resistance the invaders would look utterly foolish in expecting even a drop of water from the Indians. The advent of the nuclear age did not make any significant difference in Gandhi's position. The destructive sweep of the weaponry and war had become total, making the need for peace more urgent and enduring, yet the basic issues and dilemmas had remained the same even in the superficially changed context. So long as it was a human being who had to make the decision of using the bomb or actually firing it on people and places, even though this decision maker could not personally see the destruction wrought or hear the cries of the dying, and he had a heart beating in his breast, Gandhi believed that this perpetrator of total annihilation was not beyond the pale of sensitivity or moral transformation. And herein lay the hope for humanity, for peace.

Gandhi's activities were not designed to deal only with the "limited field of war prevention." Mere absence of war, or mechanical devices for postponing war, did not constitute peace in Gandhi's thinking. These two just did not address the basic, crucial issue of vicious exploitation of the weaker nations by the stronger. Likewise, they did not lead to tangible steps for eliminating the factors and forces that cause war. Since the stronger nations feed on the exploitation of the weaker, it was their moral responsibility to take the lead in ushering in a lasting peace. For their past sins and in retribution of their present greed, it would only be decent for the stronger, industrial nations to be sincere in their efforts at universal disarmament.

It was in this spirit that Gandhi expected a powerful nation like the United States to take the lead in bringing about disarmament, even by unilateral action. If it would not take the necessary, desirable step, the India of Gandhi's dreams was supposed to. His ardent expectations and insistent pleadings for unilateral disarmament put onerous responsibility on India, which only time will show whether India is capable of, or qualified to, shoulder it. But Gandhi is quite confident that any nation taking this bold step would assuredly pave the way for universal disarmament: such a nation would go down in history as the savior of humanity. After all, such a step is not more drastic than the one of proliferation and stockpiling of arms.

A durable and dynamic peace in Gandhi's conception, not merely a laudable objective but also a desperate necessity, can only be brought about by pure and moral means. There is such an intimate relationship, interdependence, between means and ends, according to Gandhi, that moral goals just cannot be attained by questionable means. War, with all that it entails and immoral as it is, is too degrading a means to bring forth something sublime like peace, holds Gandhi.

The history of the last two millennia--if not more--provides ample evidence of the utter futility and extravagance of war. Never has war really solved a problem--indeed, it multiplies them--or improved international relations. It originates in greed and hatred and promotes destruction and expolitation. An obsolete historical device, war is absolutely incapable of curing the ills of the world or establishing right, equality, justice, prosperity, or peace among nations. Peace gained by the nuclear deterrence is so superficial and precarious that the world now lives in constant fear of total destruction. In short, nations so far, instead of securing peace, have been moving from one kind of war and patched-up compromise or transient alleviation--rather than peace--to another. The situation is clearly intolerable; nations, pleads Gandhi, have to learn to live by loving each other, cooperating with each other.

The purity of means in bringing about the peace of Gandhi's conception, then, consists in the adoption of nonviolence in international relations, an initial step in the direction

of which is universal disarmament through unilateral disarmament, if need be, of course. India's emancipation by
means of nonviolence--and the widespread adoption of this
peaceful method by other subject peoples--would launch the
world community onto the moral path, saving the world from
an orgy of violence and infinite hatred. Nonviolence leads to
truth and a scrupulous regard for the means to bring about
desirable ends.

So the peace that would emerge from the cooperative endeavors of nations, urged Gandhi, would no longer be elu-
sive or ephemeral. Indeed it would be durable and dynamic,
for it would then be based on respect for human dignity and
national autonomy and equality. Diplomacy, the instrument of
international interaction, would have become "open" and al-
truistic. And the leaders of nations would learn to observe
the moral precepts and principles that govern the conduct of
individuals and groups.

NOTE

1. G. Ramachandran and T.K. Mahadevan, eds., Quest
for Gandhi (New Delhi: Gandhi Peace Foundation, 1970), p.
133.

BIBLIOGRAPHY

PRIMARY SOURCES

The Collected Works of Mahatma Gandhi. New Delhi: Government of India, Publications Division. To be completed in about 100 volumes, of which over 90 are in print.

Gandhi Papers. New Delhi: Gandhi National Library and Museum.

Speeches and Writings of Mahatma Gandhi. Madras: G. A. Natesen, 1933.

Gandhiji's Correspondence with the Government 1942-44. Ahmedabad: Navajivan, 1957.

Mahatma Gandhi's Correspondence with the Government 1944-47. Ahmedabad: Navajivan, 1954.

Gandhi, M.K. An Autobiography, My Experiments With Truth. Ahmedabad: Navajivan, 1959.

---. Delhi Diary. Ahmedabad: Navajivan, 1948.

---. For Pacifists. Ahmedabad: Navajivan, 1949.

---. From Yervada Mandir. Ahmedabad: Navajivan, 1935.

---. Hind Swaraj. Ahmedabad: Navajivan, 1962.

---. Non-violence in Peace and War. Vol. 1. Ahmedabad: Navajivan, 1948.

---. Non-violent Resistance. New York: Schocken Books, 1951.

---. Towards Lasting Peace. Bombay: Bhartiya Vidya Bhawan, n.d.

---. Unto This Last, A Paraphrase. Ahmedabad: Navajivan, 1956.

Gandhi Marg. Monthly of the Gandhi Peace Foundation. New Delhi.

Harijan 1933-48, 11 vols. New York: Garland, 1973.

Indian Annual Register. Volumes 1919-47. Calcutta: N. N. Mitter.

Mansergh, Nicholas, ed. The Transfer of Power 1942-7, 12 vols. London: Her Majesty's Stationery Office.

Relevant Files of the All India Congress Committee. New Delhi: Gandhi National Library and Museum.

Young India 1919-22. Madras: G. A. Ganesen, 1924.

SECONDARY SOURCES

Alexander, Horace. Social and Political Ideas of Mahatma Gandhi. New Delhi: Oxford University Press, 1949.

Amery, L.S. The Times History of War in South Africa, 7 vols. London: Sampson, Low, Marsten, 1905.

Andrews, C.F. Mahatma Gandhi's Ideas. London: Allen & Unwin, 1949.

Ashe, Geoffrey. Gandhi, A Study in Revolution. Bombay: Asia Publishing, 1968.

Azad, Abul Kalam. India Wins Freedom. Bombay: Orient Longmans, 1959.

Bainton, Ronald H. Christian Attitudes Towards War and Peace. London: Hodder & Stoughton, 1961.

Bandopadhyaya, J. Mao-tse-Tung and Gandhi. Delhi: Allied, 1973.

Bhattacharya, Buddhadeva. Evolution of the Political Philosophy of Gandhi. Calcutta: Calcutta Book House, 1969.

Bhattacharya, N.C. Gandhian Concept of State. Calcutta: M. C. Sarkar & Sons, 1957.

Bhattacharya, P. and A. Mukerji. Ruskin's Unto This Last. Calcutta: Alpha, 1969.

Birla, G.D. In the Shadow of the Mahatma. Bombay: Orient Longmans, 1955.

Biswas, S.C., ed. Gandhi, Theory and Practice, Social Impact and Contemporary Relevance. Simla: Indian Institute of Advanced Study, 1969.

Bolton, G. The Tragedy of Gandhi. London, Allen & Unwin, 1934.

Bondurant, Joan. Conquest of Violence. Berkeley: University of California Press, 1965.

Bose, Atindranath. "Gandhi: Man of the Past or of the Future." Modern Review (Calcutta, January 1957), 21-33.

Bose, Nirmal Kumar. Studies in Gandhism. Calcutta: Nishan, 1947.

---. Selections from Gandhi. Ahmedabad: Navajivan, 1950.

---. My Days With Gandhi. Calcutta: Nishan, 1953.

Brown, Judith M. Gandhi's Rise to Power. Cambridge: Cambridge University Press, 1970.

---. Gandhi and Civil Disobedience, The Mahatma in Indian Politics 1928-1934. Cambridge: Cambridge University Press, 1977.

Case, C.M. Non-violent Coercion. London: Allen & Unwin, 1923.

Catlin, George E. In the Path of the Mahatma. London: Macdonald, 1948.

Chintamani, C.Y. Indian Politics Since the Mutiny. London: Allen & Unwin, 1940.

Churchill, Winston. The Second World War. London: Cassell, 1949.

Clark, Grenwille, and Louis B. Sohn. World Peace Through World Law. Cambridge, Mass.: Harvard University Press, 1960.

Datta, D.M. The Philosophy of Mahatma Gandhi. Madison: University of Wisconsin Press, 1944.

Devanesan, Chandran S. Making of the Mahatma. Delhi: Orient Longmans, 1969.

de Kiewiet, C.W. A History of South Africa--Social and Economic. London: Oxford University Press, 1941.

de Ligt, Barthélemy. The Conquest of Violence. London: Routledge & Kegan Paul, 1937.

Desai, Mahadev. The Gita According to Gandhi. Ahmedabad: Navajivan, 1956.

Desai, V.G., trans. and ed. The Diary of Mahadev Desai. Ahmedabad: Navajivan, 1953.

Dhawan, Gopinath. The Political Philosophy of Mahatma Gandhi. Ahmedabad: Navajivan, 1951.

Diwakar, R.R. Satyagraha--Its Technique and Theory. Bombay: Hind Kitabs, 1946.

Doke, Joseph J. M. K. Gandhi, An Indian Patriot in South Africa. Varanasi: Akhil Bharat Sarv Seva Sangh Prakashan, 1959.

Erikson, Erik. Gandhi's Truth, On the Origins of Militant Non-violence. London: Faber & Faber, 1970.

Fischer, Louis. A Week With Gandhi. New York: Dull, Sloan and Pearce, 1942.

---. The Life of Mahatma Gandhi. Stuttgart: Tauchnitz, 1953.

Flint, F.S. and D.F. Tait. Lenin and Gandhi. London: C. P. Putnam, 1927.

Fox, Richard. Reinhold Niebuhr, A Biography. New York: Pantheon, 1986.

Fulop-Miller, R. Gandhi, The Holy Man. London: Putnam, 1931.

Gandhi, Prabhudas. My Childhood With Gandhi. Ahmedabad: Navajivan, 1957.

Gregg, R.B. The Power of Non-Violence. Ahmedabad: Navajivan, 1948.

Hancock, W.K. Four Studies of War and Peace in this Century. Cambridge: Cambridge University Press, n.d.

Herschberger, Guy F. War, Peace and Non-resistance. Scottdale: Herald Press, 1944.

Hingorani, Anand T., ed. Towards Lasting Peace. Bombay Bhartiya Vidya Bhawan, 1956.

Homer, A. Jack. Gandhi. Bombay: Bhartiya Vidya Bhawan, 1964.

Homes, W.H.G. The Twofold Gandhi, Hindu Monk and Revolutionary Politician. London: Mowbray, 1952.

Horowitz, J.L. War and Peace in Contemporary Social and Philosophical Theory. London: Souvenir Press, 1973.

Horsburgh, H.J.N. Non-violence and Aggression. New York: Oxford University Press, 1968.

Husain, S. Abid. The Way of Gandhi and Nehru. London: Asia Publishing, 1959.

Huxley, Aldous. Ends and Means. London: Chatto & Windus, 1957.

International Encyclopaedia of Social Sciences. London: Collier-Macmillan, 1972.

Iyer, Raghavan. The Moral and Political Thought of Mahatma Gandhi. Delhi: Oxford University Press, 1973.

Jack, H.A., ed. The Gandhi Reader. Bloomington: Indiana University Press, 1956.

James, Williams. The Moral Equivalent of War. American Association for International Conciliation, 1910.

Jones, E.S. Mahatma Gandhi. New York: Abingdon Press, 1948.

Journal of Conflict Resolution. Chicago.

Kalelkar, Kaka. Stray Glimpses of Bapu. Ahmedabad: Navajivan, 1950.

Kripalani, J.B. Gandhian Thought. Bombay: Orient Longmans, 1961.

---. Gandhi, His Life and Thought. New Delhi: Government of India, Publications Division, 1970.

Krishnamurti, Y.G. Gandhi Era in World Politics. Bombay: Popular Book Depot, 1943.

Kumar, R., ed. Essays on Gandhian Politics. Oxford: Clarendon Press, 1971.

Lester, Muriel. Entertaining Gandhi. London: Ivor Nicholson & Watson, 1932.

Lewis, John. The Case Against Pacifism. London: Allen & Unwin, 1939.

Low, D.A., ed. Soundings in Modern South Asian History. London: Weidenfeld & Nicholson, 1968.

Mahadevan, T.K. Gandhi, My Refrain, Controversial Essays 1950-1972. Bombay: Popular Prakashan, 1973.

---. Dvija. New Delhi: East-West Affiliates, 1977.

Mallik, B.R. Gandhi, A Prophecy. Oxford: Oxford University Press, 1948.

Mehrotra, S.R. "Gandhi and the British Commonwealth." India Quarterly (New Delhi, January-March 1961).

Miller, Robert. Non-violence, A Christian Interpretation. London: Allen & Unwin, 1964.

Miller, William Robert. "The Dilemma of Middle-Class Pacifism," Gandhi Marg 4:4.

Moon, Penderel. Wavell, The Viceroy's Journal. London: Oxford University Press, 1973.

Morris-Jones, W.H. "Mahatma Gandhi--Political Philosopher." Political Studies 8:1 (Oxford, 1960), 30-31.

Mukerji, Hiren. Gandhiji--A Study. Calcutta: National Book Agency, 1958.

Mukherjee, S.N., ed. The Movement for National Freedom in India. London: Chatto & Windus, 1966.

Murty, K.S. and A.C. Bouquet. Studies in Problems of Peace. Bombay: Asia Publishing, 1960.

Muste, A.J. Non-violence in an Aggressive World. New York: Harper & Brothers, 1940.

Muzumdar, Haridas T. Mahatma Gandhi, A Prophetic Voice. Ahmedabad: Navajivan, 1963.

Naess, Arne. Gandhi and the Nuclear Age. Totowa, N.J.: Bedminster Press, 1965.

Nag, Kalidas. Tolstoy and Gandhi. Patna: Pustak Bhandar, 1950.

Nanda, B.R. Mahatma Gandhi, A Biography. London: Allen & Unwin, 1959.

Nathan, O. and N. Norden. Einstein on Peace. New York: Simon & Schuster, 1960.

Nehru, Jawaharlal. An Autobiography. London: Bodley Head, 1936.

———. The Discovery of India. Calcutta: Signet Press, 1946.

———. Mahatma Gandhi. Calcutta: Signet Press, 1949.

———. Independence and After. New Delhi: Government of India, Publications Division, 1949.

———. Freedom from Fear, Reflections on the Personality and Teachings of Gandhi. New Delhi: Government of India, Publications Division, 1960.

———. A Bunch of Old Letters. London: Asia Publishing, 1960.

Niebuhr, Reinhold. Moral Man and Immoral Society. New York: Charles Scribner, 1932.

Payne, Robert. The Life and Death of Mahatma Gandhi. London: Bodley Head, 1969.

Prabhu, R.K. and U.R. Rao. The Mind of Mahatma Gandhi. London: Oxford University Press, 1946.

Prasad, Bimal. Origins of Indian Foreign Policy, Indian National Congress and World Affairs 1885-1947. Calcutta: Bookland, 1960.

Prasad, Rajendra. At the Feet of Mahatma Gandhi. Bombay: Hind Kitabs, 1955.

———. Autobiography. Bombay: Asia Publishing, 1957.

Polak, H.S.L., H.N. Brailsford, and Lord Pethick-Lawrence. Mahatma Gandhi. London: Odhams Press, 1949.

Power, Paul F. Gandhi on World Affairs. Bombay: Perennial Press, 1961.

Puri, Rashmi-Sudha. "The 'Pacifist' in Gandhi." Gandhi Marg (July 1976), 175-186.

---. "Gandhi and the Second World War." Indian Journal of Political Science 38 (January 1977), 30-55.

---. "Tolstoy and Gandhi, A Study in Pacifist Influence." Social Sciences Research Journal 7 (Chandigarh: Panjab University, November 1982), 310-325.

Pyarelal. New Horizons. Ahmedabad: Navajivan, 1959.

---. Mahatma Gandhi, The Last Phase. 2 vols. Ahmedabad: Navajivan, 1956, 1958.

---. Mahatma Gandhi, The Early Phase. 2 vols. Ahmedabad: Navajivan, 1965.

Radhakrishnan, S., ed. Mahatma Gandhi, Essays and Reflections on His Life and Work. London: Allen & Unwin, 1939.

---. Mahatma Gandhi--100 Years. New Delhi: Gandhi Peace Foundation, 1968.

Ramachandran, G. and T.K. Mahadevan, eds. Gandhi, His Relevance for Our Times. Bombay: Bhartiya Vidya Bhawan, 1964.

---, eds. Quest for Gandhi. New Delhi: Gandhi Peace Foundation, 1970.

Ray, Kshitis, ed. Gandhi Peace Memorial Number of the Visva Bharati Quarterly (Santiniketan, 1959).

Reynolds, Reginald. To Live in Mankind--A Quest for Gandhi. London: Andre Deutsch, 1951.

Robb, Peter and David Taylor, eds. Rule, Protest and Identity. London: Curzon Press, 1978.

Roberts, Adam, ed. The Strategy of Civilian Defence--Nonviolent Resistance to Aggression. London: Faber & Faber, 1967.

Rolland, Romain. Mahatma Gandhi, The man who became one

with the Universal Being. London: Swathmore Press, 1924.

Rothermund, Indira: The Philosophy of Restraint. Bombay: Popular Prakashan, 1963.

Rudolph, Susanne H. "The New Courage, an essay on Gandhi's psychology." World Politics 16 (Princeton, October 1963), 98-117.

Ruskin, John. The Crown of Wild Olive, four lectures on industry and war. London: George Allen, 1909.

Russell, Bertrand: Has Man A Future? London: Allen & Unwin, 1961.

Sahanee, Ranjee. Mr. Gandhi. New York: Macmillan, 1961.

Saxena, K.S. Gandhi Centenary Papers. 4 vols. Bhopal: Council of Oriental Research, 1972.

Seshachari, C. Gandhi and the American Scene. Bombay: Nachiketa, 1969.

Sharma, B.S. Gandhi as a Political Thinker. Allahabad: Kitab Mahal, 1956.

Sharma, Jagdish Saran. Mahatma Gandhi, A descriptive bibliography. Delhi: S. Chand, 1968.

Sharp, Gene. Gandhi Wields the Weapon of Moral Power. Ahmedabad: Navajivan, 1960.

---. "The Meaning of Non-Violence." Journal of Conflict Resolution 3 (Chicago, March 1959), 41-64.

Sheean, Vincent. Lead Kindly Light. London: Cassel, 1950.

Shraddhanand, Swami. Inside Congress. Bombay: Phoenix, 1946.

Shridharani, Krishanlal. The Mahatma and the World. New York: Duel, Sloan and Pearce, 1946.

---. War without Violence. Bombay: Bhartiya Vidya Bhawan, 1962.

Shukla, Chandra Shankar. Incidents of Gandhi's Life. Bombay: Vora, 1949.

---. Reminiscences of Gandhiji. Bombay: Vora, 1951.

---. Gandhi's View of Life. Bombay: Bhartiya Vidya Bhawan, 1954.

Sitaramayya, Pattabhi. The History of the Indian National Congress. Bombay: Padma, 1946.

Spratt, P. Gandhism--an analysis. Madras: Huxley Press, 1939.

Tagore, Rabindranath. Mahatma and the Depressed Humanity. Calcutta: Visvabharati, 1932.

Tähtinen, Unto. Ahimsa, Non-violence in Indian Tradition. London: Rider, 1976.

Tendulkar, D.G. Mahatma. 8 vols. New Delhi: Government of India, Publications Division, 1960.

Tolstoy, Leo. Kingdom of God and Peace Essays. London: Oxford University Press, 1960.

Tucker, Robert W. "Peace and War." World Politics 17 (Princeton, January 1965), 310-333.

Unnithan, T.K.N. Gandhi in Free India. Groningen, Netherlands: J. B. Wolters, 1956.

Varma, V.P. The Political Philosophy of Mahatma Gandhi and Sarvodaya. Agra: Lakshmi Narain Agarwal, 1959.

Ved, Mehta. Mahatma Gandhi and His Apostles. London: Andre Deutsch, 1977.

Watson, Francis and Maurice Brown, eds. Talking of Gandhiji. Calcutta: Orient Longmans, 1957.

Woodcock, George. Anarchism. Gretna, La./New York: Pelican, 1963.

---. Gandhi. Gretna, La./New York: Pelican, 1969.

Woodruff, Philip. The Men Who Ruled India, The Masters. London: Jonathan Cape, 1959.

---. The Men Who Ruled India, The Guardians. London: Jonathan Cape, 1959.

Yajnik, Indulal K. Gandhi As I Know Him. Delhi: Danish Mahal, 1943.

INDEX

ABOUT THE AUTHOR

RASHMI-SUDHA PURI (nee Kaura) holds a First-Class M.A. degree in Political Science, a Ph.D. from the Panjab University, and has, for over a decade now, concentrated her interest on Gandhi. She has published in scholarly journals like Gandhi Marg and the Indian Journal of Political Science. Presently on the faculty of the Department of Gandhian Studies at the Panjab University Chandigarh in India, Dr. Puri has taught in West Germany and offered courses on Gandhi at the University of Idaho and the Portland State University.